Anderson Peretto
Ivaniza L. L. Cabral

Ribeirão Grande: Natural Dynamics and its Production of Geographic Space

Anderson Peretto
Ivaniza L. L. Cabral

Ribeirão Grande: Natural Dynamics and its Production of Geographic Space

ScienciaScripts

Imprint

Any brand names and product names mentioned in this book are subject to trademark, brand or patent protection and are trademarks or registered trademarks of their respective holders. The use of brand names, product names, common names, trade names, product descriptions etc. even without a particular marking in this work is in no way to be construed to mean that such names may be regarded as unrestricted in respect of trademark and brand protection legislation and could thus be used by anyone.

Cover image: www.ingimage.com

This book is a translation from the original published under ISBN 978-3-639-68941-9.

Publisher:
Sciencia Scripts
is a trademark of
Dodo Books Indian Ocean Ltd. and OmniScriptum S.R.L publishing group

120 High Road, East Finchley, London, N2 9ED, United Kingdom
Str. Armeneasca 28/1, office 1, Chisinau MD-2012, Republic of Moldova, Europe
Printed at: see last page
ISBN: 978-620-8-07542-2

DEDICATION
> *I dedicate it to the great love of my life who has supported me so much in the completion of this work, my beloved wife Katherine Pinoti.*

ACKNOWLEDGEMENTS

I would like to thank God for protecting me and guiding my steps and thoughts during this academic journey that I have been allowed to travel and complete and for which I have received a good reward.

I would like to thank Maria José de Toledo for all her support and encouragement so that I would never give up in the face of the difficulties I would encounter along the way.

aaI would like to thank my supervisor, Prof. Dr Ivaniza de L. L. Cabral, for her understanding and the time she gave me during my orientation phase, during which I was able to learn a lot through her guidance.

My special thanks go to my great friend Roberto Nunes Viancone Souto, who always opened the doors of his house to me and helped me a lot when I needed it most, and who proved to be a very generous friend, a person who is always ready to help his neighbour. I would also like to thank all my fellow students from my Master's programme in 2010.

I would like to thank all the professors of the Postgraduate Geography Programme at the Institute of Human and Social Sciences of the Federal University of Mato Grosso. And to Professors Dr Célia Alves de Souza and Maria Aparecida Pereira Pierangeli from the Environmental Sciences Postgraduate Programme at the State University of Mato Grosso, where I studied their subjects, for all the knowledge they imparted to me in class and also in the various discussions that these professors held.

I would like to thank the Foundation for Research Promotion of the State of Mato Grosso for the grant I received to carry out this work.

ABSTRACT

The production of the geographical area in the highlands of Mato Grosso is directly related to its natural potential, which in the current situation accounts for a large part of the economic activities. In this context, the watershed of the Ribeirao Grande, a left tributary of the Teles Pires in the south of the municipality of Sorriso, in the north of the state of Mato Grosso (Brazil), was defined as the working area. The main objectives were the environmental analysis of the basin and the verification of the relationships between the morphometric aspects of land use and these aspects with geographic space production. The occupation and transformation of the geographic space of the municipality of Sorriso was contextualised through the characterisation of the natural elements of the basin and the theoretical revision of the themes and objects used. The work is based on geosystem theory and uses various geoprocessing methods and tools, in particular Spring 5.7.1 and Arcgis 9.3 software, to develop a detailed basis for the drainage network and land use mapping. Background sediment and landworks analysis was used to gain knowledge of the working area, validation of land use data, description and analysis of relevant elements for the working theme. [2]With an area of 415.3 km and a low density of drainage and watercourses, the Ribeirao Grande basin has a fourth-order fluvial hierarchy, a flat relief with tabular shapes, 79% agricultural use and 20% protected areas, which present several environmental problems due to their land use and land production.

Keywords: Morphometric analysis; Land use; Ribeirao Grande basin; Geographic spatial production.

INTRODUCTION

In recent decades, environmental issues, including in river basins, have become one of the most important points of discussion in all areas of society. In general, society realises how important it is to use natural resources properly and demands that all areas of production use techniques that cause fewer problems for the environment.

Academic work has an even greater responsibility, namely to scientifically review the real situation of river basins and to propose measures to mitigate the problems that may exist with the help of science and various techniques.

According to Cristofoletti (1980), river catchments consist of a network of drainage channels that are interconnected and form a catchment, defined as the area drained by a particular river or river system. The amount of water entering the river courses depends on the size of the catchment, the total amount of precipitation and the drought regime, which allows for losses through evaporation and infiltration.

[0]The introduction of the river basin as an environmental management and planning unit is a trend in many countries, which has intensified in Brazil with the adoption of the National Water Resources Policy (Law No. 9433 of 1997). This law introduced river basin committees responsible for the decentralised and participatory management of water resources in the river basins.The occupation of the Brazilian Cerrado intensified from the 1960s and came about through the occupation projects of the military government, whose motto was "*occupy in order not to surrender*", which enabled national integration and raised funds for the construction of highways linking the Centre-West and Amazon regions with the other Brazilian regions. It was the federal highways that enabled effective occupation and also allowed better circulation of the population and the flow of production in these regions (SANTOS et al, As part of the integration of the Cerrado, the land in the Ribeirao Grande catchment area was developed, supported by advances in agricultural production techniques such as the correction of soil acidity, the use of fertilisers, pest control and the intensive use of mechanisation.A large part of the state of Mato Grosso is occupied by the Cerrado biome, the biological unit in which the Ribeirao Grande catchment area is located, in the municipality of Sorriso, which has an area of soils suitable for mechanised agriculture, predominantly of the latosols category, typical of the flat relief units that form part of the dissected Parecis plateau. Together with the climate with two clearly defined seasons, one dry and one rainy, these factors provide good conditions for the excellent productivity of the crops of this municipality.the areas with the highest agricultural and livestock production in the state of Mato Grosso are located in the catchment area of the Teles Pires River, which shows that it is one of the most important hydrographic units in the state from an economic point of view. The hydrographic basin of Ribeirao Grande is one of the sub-basins of the middle sector of the Teles Pires River basin, an area destined to carry out the predominant agricultural economic activities in the region, which emphasises the importance of carrying out this study.

To this end, the work focused on the general objective of carrying out an environmental analysis of the Ribeirao Grande catchment and relating the morphometric aspects and the type of land use to the creation of the geographical space.

Knowing the dynamics of these elements is key to proposing intervention measures and spatial planning, given the importance of this area for the state.

From this point of view, the research included the following specific objectives: (I) characterisation of the natural elements, geology, pedology, geomorphology and climate; (II) analysis of the characteristics of the drainage network patterns in the catchment, using morphometric parameters; (III) mapping related to the exploitation theme; (IV) analysis of human activities in shaping the geographical space of the Ribeirao Grande catchment.The work is divided into seven parts, presenting the processes carried out to obtain the materials, analyses and results of the research. In the first part, the characteristics of the process of occupation and transformation of the geographical space of the municipality of Sorriso, where the study area is located, are presented. In the second part, the physiographic aspects of the Ribeirao Grande river basin are described. In the third part, the bibliographical study was carried out, in which the contextualisation and justification of the topics dealt with was developed. The methodological approach is presented in the fourth section, which presents (I) the theoretical and methodological foundations and concepts of the work, (II) the technical procedures used to process the spatial data and the formulas and parameters used in the morphometric analysis, and (III) the laboratory procedures and data collection in the field.The fifth section presents the results and discussion of the data analysis and morphometric indices, physical parameters, vegetation description and analysis, mapping and characterisation of land use types and, finally, the creation of the geographical space and its effects. The sixth section presents the conclusions drawn from the studies carried out and the recommendations for the Ribeirao Grande catchment and the region; finally, the seventh section contains the bibliographic references used to support the discussions in this thesis.

This work is the first to be carried out in this area for this purpose. Cabral (2011) carried out work in this region to assess the degree of degradation caused by the practice of artificial drainage associated with the cultivation of maize and soya in the wide and very wide interfluves that make up the lands of the southern sector of the municipality of Sorriso-MT.

PROCESS OF OCCUPATION AND TRANSFORMATION OF THE SORRISO
    GEOGRAPHICAL AREA

The colonisation of the municipality of Sorriso began at the time of the expansion of the Brazilian agricultural frontier towards the Amazon. Sorriso received incentives from the military governments in the late 1970s for its colonisation and settlement, as it is located in the rainforest known as the legal Amazon.

Originally, the settlement received immigrants from the states of Paraná, Santa Catarina and Rio Grande do Sul, more specifically from the Passo Fundo region. These immigrants were brought by the Colonizadora Feliz, whose coloniser was Claudino Francio, who had acquired a large area in the central north of Mato Grosso. The first families to settle in the area of the current town of Sorriso were the Silva and Santos families (FERREIRA, 2008).

In 1980, the small agro-village in Mato Grosso was elevated to a district belonging to the municipality of Nobres. Later, in 1982, a sub-municipality was created in the Sorriso district with Mr Genuino Spenassatto as sub-mayor. The municipality was emancipated by the Legislative Assembly of the State of Mato Grosso on 13 May 1986 (FERREIRA, 2008). Shortly after emancipation, the population of Sorriso exploded, helped by the asphalting of the BR 163 motorway.

The construction of the BR 163 motorway has also led to a reduction in the cost of transporting grain, thus enabling the region's agricultural potential to be expanded by improving the competitiveness of agricultural products from the northern part of the state of Mato Grosso on the national and international market.

In 1979, the municipality of Sorriso was a village, as can be seen in Figure 01, with only a few houses along the BR 163 motorway.

Source: http://www.achetudoeregiao.com.br/MT/sorriso/historia.htm. Retrieved 10/06/2011.

According to Ferreira (2008), the name Sorriso has its origins in the nature of the events, which is linked to the pioneering spirit of the immigrants. According to the accounts of the first inhabitants brought by the Feliz colonisation company, they believed that they should always keep a smile on their faces in order to overcome the work and the great difficulties.

In just over two decades, Sorriso has developed from a simple village in the middle of the vast Cerrado into the municipality that is now Brazil's largest soya producer.

Sorriso is an important regional city centre with a high degree of modernity. It is experiencing rapid urban growth, with various signs of the phenomenon of urban verticalisation, with the construction of residential buildings, as shown in Figure 02, A and B.

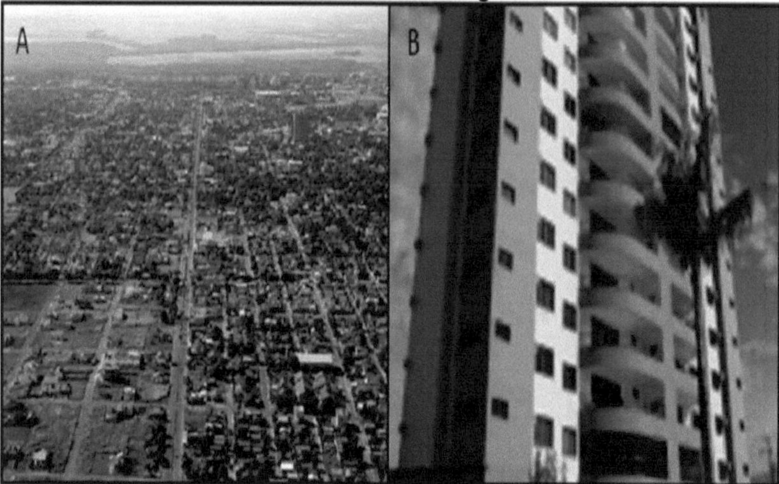

Source: http://www.sorriso.mt.gov.br. Retrieved 24/06/2011.

Today, the town of Sorriso has a good range of services, several branches of multinational companies linked to the agricultural industry, companies that process agricultural products, as well as equipment and machinery, a shopping centre, a regional hospital, specialised clinics, good supermarkets and other businesses of a high standard.

The town of Sorriso is growing due to the large number of high standard buildings and the presence of several properties with modern architectural design. [2]Current figures show that the municipality has 66,521 inhabitants and an area of 9,330 km (IBGE, 2010 census).

Sorriso was separated from the municipalities of Nobres, Diamantino and Sinop. It is located on the BR 163 motorway and has the following boundaries: with the municipality of Sinop, Vera (natural boundary Rio Celeste and Teles Pires), Lucas do Rio Verde, Nova Mutum and Santa Rita do Trivelato (natural boundary Rio Verde and Teles Pires), Nova Ubiratá, Tapurah and Ipiranga do Norte (natural boundary Rio Verde). Source: http://www.sorriso.mt.gov.br, accessed on 24/06/2011.The municipality of Sorriso has a high technological level of agricultural production, which is reflected in the high production volume per hectare. It is characterised by the use of state-of-the-art technology, making it one of the most modern agricultural centres in the world. The high technological level and the production processes used in planting can be seen in Figure 03, where soya is harvested and then the second crop of the year is planted, which could be maize or sorghum, demonstrating the intensification of the agricultural production chain in the state of Mato Grosso at all levels.

Source: http://www.mochileiro.tur.br/sorriso.htm. Retrieved 24/06/2011.

When analysing the economic situation of the state of Mato Grosso with a focus on the primary sector, the state currently stands out as Brazil's largest soy producer. According to Moreno (2005), who analysed data from the government of Mato Grosso, the municipalities with the highest soy production are concentrated in the Alto Teles Pires microregion, where 32.4 % of Mato Grosso's total production is produced. As far as livestock farming is concerned, the municipalities with the largest livestock populations are located in the north of Mato Grosso and are also part of the Teles Pires river basin, where 38.2 % of the state's total livestock is kept.

THE RIBEIRAO GRANDE RIVER BASIN AND ITS CONTEXT

°The Ribeirao Grande basin is located between coordinates 12 30' 12" and 12o 53' 41" south latitude and 55o 57' 54" and 55o 46' 50" west longitude from Greenwich, in the southern sector of the municipality of Sorriso-MT, 412 km from the capital Cuiabá (see Figure 04).

Figure 04 - Map of the municipality of Sorriso-MT and the hydrographic unit of Ribeirao Grande

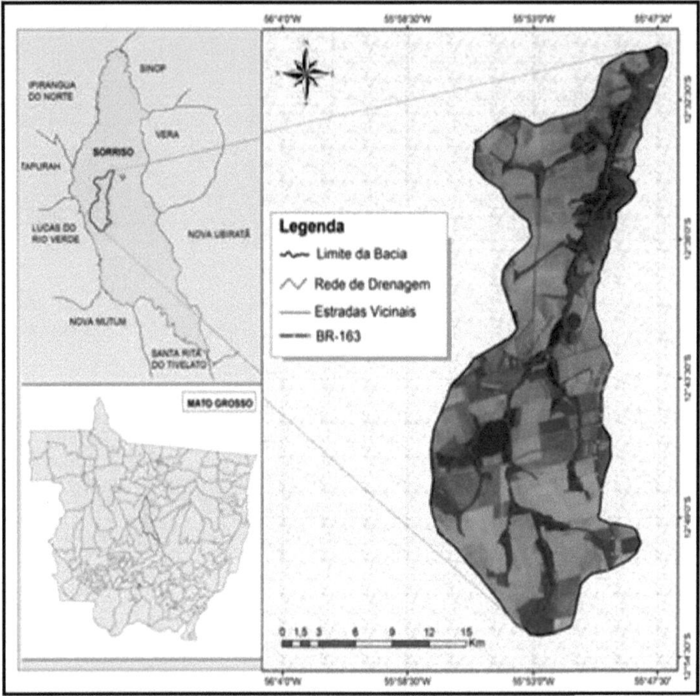

Source: Based on the information in the maps of the political subdivision and the hydrographic base of the state of Mato Grosso (SEPLAN/MT, 2007).
Elaboration: PERETTO, A. (2011)

The Ribeirao Grande is one of the tributaries on the left bank of the middle reaches of the Teles Pires river. The catchment area of the Teles Pires is part of the northern macro-region of the state of Mato Grosso, also known as the Legal Amazon.

Its main tributaries are the Peixoto de Azevedo, Cristalino, Verde, Do Lira and Moroco rivers. At a regional level, the Teles Pires River and its tributaries are of great importance to the state, as these drainage networks are responsible for supplying water to several municipalities such as Sorriso, Lucas do Rio Verde, Nova Mutum, Sinop, Colider and Alta Floresta.

In terms of land use in the region of the middle reaches of the Teles Pires, activities related to agribusiness, the flagship of the regional economy, stand out. According to Santos et al. (2004), the mechanisation and implementation of agriculture in Sorriso and in the Teles Pires river basin began in the 1970s with an extensive occupation of the Cerrado by means related to the expansion of the agricultural frontier, which led to the inclusion of areas of this vegetation cover in the national scenario for agricultural production with a high degree of mechanisation.

The centralised federal government intervened in the process of appropriating the Cerrado from 1970 onwards with the implementation of the Second National Development Plan (II PND). The federal government also launched the Cerrado Development Programme (POLOCENTRO - 1975/1979), which created numerous mechanisms to make businesses viable in the Cerrado regions (SANTOS, 2004).

The process of expanding the agricultural boundaries of the Cerrado also attracted international investment - international interest in the region grew and was favoured by POLOCENTRO. The Japanese were the first to invest foreign capital in the Brazilian Cerrado biome under the Japanese-Brazilian agreement (SANTOS, 2004).

CHAPTER 2 PHYSIOGRAPHIC ASPECTS

This section contains regional data on the climatic conditions, geology, geomorphology, soils, hydrographic network and vegetation cover, as it is important to understand the natural resources in the Ribeirao Grande catchment area in the municipality of Sorriso.

CLIMATE

As already mentioned, the municipality of Sorriso, where the study area and its plots are located, does not have a uniform climate. It is therefore possible to distinguish a tropical climate type with contrasting seasons, i.e. number 2 (two) according to the classification of Durand-Dastés (1968) for the major climatic groups, modified by Estienne and Godard (1970) and presented by Tardy (1986).

In this way, the year is divided into two seasons characterised by the amount of precipitation: a dry season and a rainy season. With this seasonal variation, there are six hot months with fluctuations between extreme heat and extreme cold and six rainy months (BITTENCOURT ROSA et al., 2002).

From an ecological point of view, the municipality of Sorriso belongs to the AWI type according to the Koppen classification - a humid tropical climate with a pronounced dry season (winter/summer), with an average temperature difference between the warmest month (October) of around 37°C and the coldest (June) of around 15°C. The average annual rainfall is around 2,000 mm and the average annual temperature is 26°C. The average annual rainfall is around 2,000 mm and the average annual temperature is 26°C.

The rainy season usually begins in September and lasts until April. The summer months from December to March are characterised by a significant increase in regional precipitation, 80 % of which falls during this period (BITTENCOURT ROSA et al., 2002).

REGIONAL GEOLOGY

Geologically, the study region consists of depositional episodes that extend from the Upper Cretaceous to the Recent Alluvium and correspond to the lithostratigraphic units (Mendes, 1996): Parecis Group (Salto das Nuvens and Upper Cretaceous Utiariti Formations); Detrital-Lateritic Covers related to the Tertiary and Recent Alluvium (BARROS *et al.* 1982, SCHOBBENHAUS *et al.* 1984, WESKA, 1996, WESKA *et al.* 1996, BITTENCOURT ROSA *et al.* 2002, BITTENCOURT ROSA, 2006, 2007 and WESKA, 2006).

Parecis Group

The name Parecis Group was proposed by Barros et al. (1982) in view of the geographical and geological extent of this unit and its characteristics. It corresponds to the lithological group that delimits the basin of the Upper Paraguay River and the Amazon River.

Recently, Weska (2006) published a summary of his studies on the Upper Cretaceous in the state of Mato Grosso. This author analysed the lithostratigraphic unit of the Parecis Group in comparison to the Bauru Group. This is because it is the oldest on the geological scale of the state. Weska (2006) presents a new stratigraphic column in which only the Parecis Group occurs, with the following lithological composition, distributed from bottom to top as follows: Formagóes Paredao Grande, Salto das Nuvens, Cachoeira do Bom Jardim and Utiariti.

It should be noted that the Bauru Group was first studied and recognised by Gonzaga de Campos (1905), who was cited by Almeida (1948) for the sandy limestone sediments of the Paraná River plateau in the state of Sao Paulo and referred to it as the *"Baurú Group"*. Soares et al. (1980) categorised this unit in the state of Sao Paulo as a "group".

The Bauru group was confirmed by Weska (1987) during investigations in the Chapada dos Guimaraes in the state of Mato Grosso.

According to Oliveira (1992) and Oliveira et al. (1992), these units were correlated with those whose type segments were described by Weska (1987) and later recognised by Araújo et al. (1991) in the municipalities of Dom Aquino and Poxoréu.

Salto das Nuvens formation

This Upper Cretaceous lithostratigraphic unit was first studied by Barros et al. (1982) and, according to Weska (2006), is correlated with the units defined in the regions of Chapada dos Guimaraes (1987) and Dom Aquino and Poxoréu (1991): Chapada dos Guimaraes, by Weska (1987), and in Dom Aquino and Poxoréu, by Pisani and Arrais (1991).

The Salto das Nuvens formation also outcrops in fault contact with the Paredao Grande and Botucatu formations and combines deposits on the slope edge in the form of polymictic conglomerates, 90 % of which consist of volcanic components from the Paredao Grande formation. The remaining 10 % consists of pebbles and shells from the Raizama Formation (Alto Paraguay Group), Aquidauana, Palermo and Botucatu. These conglomerates show an inverted gradation, with several cycles of alternation between conglomerates and lenses of clay and clayey sandstones, sometimes with cross-stratification (WESKA et al., 1996).

Utiariti formation

This lithostratigraphic unit, which is also dated to the Upper Cretaceous, is the top of the Parecis Group. According to Weska (2006), it also correlates with the sections defined by Weska (1987) in the Chapada dos Guimaraes and in Dom Aquino and Poxoréu, whose type sections were described by Weska (1987) as Cambambe facies in the Morro do Cambambe region in the Chapada dos Guimaraes, later recognised in the municipalities of Dom Aquino and Poxoréu by Maciel and Ribeiro (1991), Araujo et al. (1991) and Pisani and Arrais (1991), when it was elevated to the Cambambe Formation class by Weska et al. (1996) and is now re-examined by Weska (2006) as the Utiariti Formation, which belongs to the top of the Parecis Group.

The Utiariti Formation consists lithologically of cyclic, oligomitic basal conglomerates with pebbles and rare sandstone and quartz clasts with a sandy matrix and siliceous cement. In the middle to upper sections, conglomeratic sandstones, siltstones, sandstones and clayey siltstones, sometimes interbedded with lenses of micro-conglomerates, predominate. The average thickness of this package is 100 metres (WESKA 2006).

Geology of Ribeirao Grande

[2]The study area is geologically arranged over the Detrital-Lateritic Cover edaphostratigraphic unit, which covers a large area of SD 21 mapped during the implementation of the research activities of the RADAMBRASIL project, approximately 80,000 km. The stratigraphic structure consists of a lithological unit arranged in erosional unconformity that covers the entire study area.

Detrital-lateritic cover

Detrital-lateritic coverings or laterites are crusts of iron oxide of a dark reddish to brownish-yellow colour, which occur solid or in oolites and pisolites. They can occur irregularly. The massive forms are characterised by regular crusts with a thickness of a few centimetres, from 0.5 to 5 cm (BARROS et al., 1982).

The oolitic and pisolitic laterites contain iron oxide secretion nodules, and the irregular shapes also represent irregular crustal planes. In some places, quartz grains are scattered in the laterite crust.

According to Barros et al. (1982), the detrital-lateritic overburden is an edapho-stratigraphic unit whose formation is related to pedogenetic processes on flattened surfaces in tropical climatic regions.

The development of this unit only takes place where climatic conditions alternate between dry and rainy seasons in the region, i.e. with rainfall of over 950 mm per year, an average temperature of 25 °C and a flat topography with a slope of 8 to 10 % (WESCA, 2006).

Latest allocations

These alluvial deposits are a group of sediments found on the banks of rivers that are periodically affected by floods and are inundated most of the time. They are therefore the product of soil transport by water currents that form successive layers in different horizons, the deposits often consisting of sands, clays, silts, ferruginous concretions, etc. (BARROS etal., 1982).

REGIONAL GEOMORPHOLOGY

From a geomorphological point of view, the region in which the catchment area of Fazenda Sao Roque II and its surroundings are located belongs, according to Ross and Santos (1982), to the catchment area of the Middle Arino in the geomorphological unit known as the Parecis Plateau.

The Parecis Plateau was first studied by Derby (1895) and then by Melo et al. (1978) as part of the geomorphological mapping of sheet SC.20/Porto Velho for the RADAMBRASIL project. These authors had defined this unit as sedimentary in these studies, but according to Melo and Franco (1980) they mapped sheet SD.20/Guaporé with the extension of the project.

This geomorphological unit is one of the most extensive and continuous on sheet SD.^{2} 21/Cuiabá at a scale of 1:1,000,000 of the RADAMBRASIL project. It occupies almost the entire northern part of the sheet and covers an area of 63.497 km (MELO and FRANCO, 1980). According to the aforementioned authors, they correspond to the sub-unit of the Parecis Plateau, labelled Parecis Dissected Plateau by Melo et al. (1978) and located on sheets SD.20/Guaporé and SD.21/Cuiabá by Ross and Santos (1982), which are also depicted at the same scale by the aforementioned project.

The Parecis plateau is partly homogeneous and has predominantly tabular, jagged shapes, the altitude of which varies between 350 and 420 metres from east to west. The degree of dissection varies from east to west. These different aspects are related to the equally variable rock composition, so that this unit can be divided into two relief models, with the confluence of the Arinos River and the Teles Pires River as the separating element (ROSS and SANTOS, 1982).

One of the compartmentalisations, which begins on the right bank of the river, extends eastwards, outside the limits of the Teles Pires basin, towards the

Xingu valley, and the other, which moves westwards from this river, also outside the limits of the basin, already in the area where the Aripuana basin predominates (ROSS and SANTOS, 1982).

The Parecis Plateau is divided into the dissected Parecis Plateau and the Parecis Plateau.

Chapada dos Parecis

The Parecis Dissected Plateau hosts the Parecis Plateau, which extends over a large flattened area with elevations of up to 550 metres covered by a Tertiary detrital laterite deposit (ROSS and SANTOS, 1982).

As a result of the resumption of erosive processes, the Chapada dos Parecis has retreated by dissecting into erosive amphitheatres that generally converge into broad, deep valleys bordered by abrupt ridges and escarpments (ROSS and SANTOS, 1982).

Geomorphology of Ribeirao Grande

The geomorphology of the study area was described by Ross and Santos (1982) in the RADAMBRASIL report. The Ribeirao Grande hydrographic unit is located in the macrogeomorphological unit of the Parecis Plateau, which was divided into two morphological subunits, namely the Chapada dos Parecis itself and the dissected Parecis Plateau, due to the peculiarities associated with the original surface and its reshaping. The Ribeirao Grande hydrographic unit is located in the macrogeomorphological unit of the Parecis Plateau, which has been divided into two morphological subunits, namely the Chapada dos Parecis proper and the unit of the dissected Parecis Plateau, in which the entire area of the hydrographic unit is located, due to peculiarities related to the original surface and its transformation.

Prepared Parecis plateau

It corresponds to the geomorphological unit that comprises an important area of plateaus distributed over Palaeozoic and Cenozoic terrain, forming the watershed of the Platine and Amazon basins, where the relief is dissected, with large tabular forms, residual elevations with flat peaks bordered by ridges and escarpments that represent stepped structural planes (BITTENCOURT ROSA et al., 2002).

The dissected Parecis plateau is bordered to the west by the Guaporé hollows over convex structural escarpments; to the south, in an east-west direction, it borders the Cuiabana and Alto Paraguay hollows and the Paranatinga plateau. To the north, it borders the Goiás plateau. The Parecis Plateau is divided into two sub-units, (I) Chapada dos Parecis and Planalto Dissecado dos Parecis, where the Corrego Grande basin is located in the municipality of Sorriso.

According to Ross and Santos (1982), the main characteristic of the Parecis dissected plateau is the continuity and homogeneity of the relief, where tabular, dissected forms predominate. However, the dissection of the relief occurs from east to west, which is related to lithological differences, and in the south a differentiation of the relief can be observed, which is also influenced by lithological variations.

REGIONAL SOILS

The regional soils are represented by: Concretionary soils, Latosols, Cambisols, Regolitic Neosols, Litholiths and Quartzars, Argisols and Organosols (OLIVEIRA et al., 1982; EMBRAPA, 1999, 2006; BITTENCOURT ROSA et al.,

2002; MOREIRA and VASCONCELOS, 2007).
Concrete soils (plinthosols)

The occurrence of these soils is linked to the rocks of the Salto das Nuvens formation in the region.

These soils are also only sparsely present in the more dissected parts of the relief, which are based on the lithologies of the Tertiary detrital-lateritic overburden.
Latosole

These are soils with a weakly developed A1 horizon no more than 20 cm thick, generally with a low organic matter content, with a structure, texture and colouring that vary from site to site (BRAUN, 1962).

Chemically, they are characterised by an acidic pH value that fluctuates between 3.5 and 5.0. Alkalis such as calcium, potassium and magnesium are not very representative in these soils. The organic matter content is low, as is the phosphorus content, but the amount of aluminium and iron oxides exceeds the desilicon content (BRAUN, 1962).
Cambisols

Cambisols have an A horizon, which is usually of medium type and lies above an incipient B horizon. They are generally shallow, with a constant presence of the O, A, E, B and C horizons. These soil types are often found in areas where the Parecis group predominates, when they have low aluminium saturation and can be eutrophic or dystrophic. According to Oliveira et al. (1982), when aluminium saturation is high (albic), they occur predominantly in expanses and occupy some sectors of the Parecis plateau associated with the rocks of the Salto das Nuvens formation, forming horizons with a gravelly texture.

They are stony and occur in areas with hilly to partially hilly relief, which, combined with the low natural fertility, means that this soil type is of little interest for agriculture and is generally used for grazing, while it is widely used in construction (OLIVEIRA et al., 1982).
Regolithic neosols and lithic neosols

Regolithic neosols and litholithic neosols develop on the deeply weathered rocks found in and around the area covered by this research. There was not enough time for the soils to form, as these areas are currently being cut up by intense erosion. The most common lithic neosols in the region are conglomerates and sandstones (KER et al., 1990).

These soils are shallow and have a low organic matter content. The pH value is acidic and low, around 3.7. The mobility of iron and aluminium oxides is low and the leaching of SiO_2 is low. The phosphorus content is low (KERetal., 1990).Quartzarenic neosols

This type of soil belongs to the class of sand-quartz soils that develop from sandstones or unconsolidated sand-quartz sediments of the Parecis group. It is poorly developed, with the continuity of the A and C horizons, with a low capacity to retain water and cations, and is also particularly unsaturated. These are the predominant strata in the region (OLIVEIRA et al., 1982).

The quartzarenic neosols (formerly quartz sands) were described by Ker et al. (1990) as soils with a simple structure in which there is no coherence between the structural units due to the absence of aggregation colloids (organic matter, oxides and clay). As such, they are highly susceptible to erosion, and in the areas where they predominate, gullies and flooding are not uncommon, especially when human intervention occurs.

Erosion processes develop quite easily in these soils and their control requires very expensive procedures, which, in combination with chemical and physical factors, makes agriculture more difficult (KER et al., 1990).

ArgisolsThis type of soil is characterised by the fact that the sediments give rise to poor soils with an exchangeable aluminium content, which are particularly well drained and well leached and develop from materials of various origins in the areas with the greatest relief. They can generally be eutrophic, dystrophic or alkaline. The characteristic feature is the presence of clay in the deeper layers (KER et al., 1990).Due to the textural gradient, argissolos can pose a serious risk for erosion processes because of the different water infiltration through the profile, i.e. faster in the A horizon, which is sandier, than in the B horizon, which is usually loamier (KER et al., 1990).Soils in the Ribeirao Grande catchmentThe pedological units that predominate in the Ribeirao Grande catchment are mainly represented by the categories of red latosols, yellow latosols and fluvial neosols, represented by the arganosols, which are located along the axes of the main rivers. According to EMBRAPA (1999), the latosols are characterised by a high rate of development, reflected in the thick profiles and the latosolic B horizon, made possible by the conditions associated with the hydric functioning of their waters and favoured by the hilly and flat topography. In general, latosols are very weathered and deep soils with good drainage. They are characterised by a high degree of homogeneity in their properties across the entire profile. The mineral composition of the clay fraction is kaolinitic or kaolinitic-oxidic, which is expressed in low Ki values and the absence of easily weatherable primary minerals. They are present in large parts of the Brazilian territory, occurring in all regions and differing mainly in their colour and iron oxide content (IBGE, 1999).Latossolo VermelhoThe Latossolos Vermelhos have good depth, good drainage and low natural fertility. They are found practically all over Brazil, but are not very common in the north-eastern states and in Rio Grande do Sul. When they have a loamy texture, they are often used for mechanised cereal cultivation, and when they have a medium texture, they are mainly used for pasture (EMBRAPA, 1999). Yellow latosolsYellow latosols are deep soils with a yellowish colour, very homogeneous profiles, good drainage, but low natural fertility. These soils are found in large parts of the lower and middle Amazon region and in the coastal wetlands (tabuleiros) (EMBRAPA, 1999). In the Ribeirao Grande, they are found in the areas with wide and very wide river courses.Organosols

These soils are typical of the floodplains of the Ribeirao Grande basin, where alluvial sedimentation has only recently occurred. They are well drained and correspond to a high proportion of alluvial soil. Their characteristic A horizon is no more than 2 metres thick. The pH value of these soils is variable, it is acidic and varies between 3.7 and 5.2. The alkalis Na, Ca and Mg are low, as is the phosphorus. Locally, these soils are sandy, contain some clay and silt and are hydromorphised (KER et al., 1990).

HYDROGRAPHY

The local hydrographic network belongs to the Teles Pires catchment, which is represented by the hydrographic basins of its tributaries and the Ribeirao Grande catchment, as shown in Figure 05.

Figure 05 - Partial view of the Ribeirao Grande sewer

Source: PERETTO, A. (2011).

[a]The Ribeirao Grande has first order drainage channels with an undefined bed, while from the second order channels onwards the drainage channels have been filled in. Located at 12° 32' 29.3" south latitude and 55° 48' 38.2" west longitude, it has riparian vegetation with medium to large trees, as evidenced by the characteristics of the drainage channel in the lower reaches of the basin.

Any scientific work must have a certain prior knowledge as a basis, which serves as a guide for the development of the work to be carried out, because if you have no prior knowledge, you have no theoretical basis and no prediction of the results you will obtain.

The works of the authors cited in this chapter of the bibliographical overview served as a basis and support for the development of this thesis.

RIVER AREAS

Watersheds consist of slopes, springs and rivers that have the same starting point: the main river, and thus form the drainage network. Here, natural factors such as geology, geomorphology, soil science, vegetation and climate interact with anthropogenic activities (PERETTO, 2010).

According to Tucci (1995), watersheds correspond to a natural catchment area for rainwater in which all the water flowing into the unit has a single outlet, the so-called exultorium. They essentially consist of a series of surface slopes and a drainage network made up of watercourses that flow into the same outlet. The slopes form the flow system of the entire catchment and can be interpreted as producing the water for the surface runoff and forming the springs that generate the outflow to the drainage network responsible for transporting it to the outlet of the catchment.

A catchment is an area on the earth's surface in which water, sediment and dissolved substances flow to a common outlet or to a specific point in a river channel. Its boundary is called the watershed or catchment area (NETO, 2001).

In practical terms, Guerra and Cunha (1996) have pointed out that hydrographic catchments are excellent units for the management of natural and social elements due to their integrative nature, since through this perspective of analysis it is possible to follow the changes introduced by man in the territory of a catchment and the corresponding reactions of nature. The same article points out that in the more developed countries, the river basin is used as a planning and management unit that unites the different interests of water use and ensures its quantitative and qualitative character.

[0] With the adoption of the Law *on National Water Resources Policy* (Law No. 9,433 of 1997), which provides for the establishment of river basin committees and agencies and the participation of civil society organisations in environmental planning and the preparation of master plans for the use of river basins. These measures help to increase interest in river basin studies.

In his study on the alteration of the drainage network in micro-catchments, Collares (2000) considered that the study of drainage network alteration diagnoses in river basins at the regional level is a good indicator for the promotion of environmental zoning measures in river basins. The drainage network is susceptible to various forms of utilisation, such as urban, agricultural and mining activities, to name but a few. All these factors can lead to changes in structure and form or even cause losses or the creation of new channels, changing the dynamics of surface water runoff from river catchments.

FLUVIAL GEOMORPHOLOGY

River geomorphology studies the forms that rivers take in their course and analyses the elements that influence the different types of channels in nature.

According to Cunha (2000), fluvial geomorphology is concerned with watercourses and river basins. In these studies, fluvial processes and the forms resulting from water runoff are evaluated. In the study of river basins, features such as geological and geomorphological aspects, hydrology, climatic conditions, biota and human settlement of the land are taken into account, which together determine the hydrological system.

Cunha and Guerra (1996) report that geomorphology is of paramount importance in environmental impact studies. These authors point out that the breakdown of natural harmony caused by environmental degradation in a catchment or part of it can lead to the development of a vogoroca or other types of erosive processes. It is of utmost importance to know the dynamics of the evolution of the degradation processes that occur and unbalance the natural harmony of the environment.

When investigating river courses or channels, the focus is on fluvial processes and the resulting forms of water flow. When investigating river catchment areas, the most important features that determine the hydrological system are analysed. These features can be related to geological aspects, landforms, hydrological and climatic characteristics, biota and land use (CUNHA, 2001).

TAIL TYPES

According to Cunha (2001), the channel type is the physiognomy that the river exhibits along its longitudinal course and can be described as a straight, anastomosing and meandering channel. This channel geometry results from the adaptation of the channel to its cross-section and reflects the interrelationship between liquid discharge, sediment, gradient, channel width and depth, flow velocity and bed roughness. From this it can be concluded that the channel types result from the interrelationships between the variables present in the channels.

In a single river basin, channels can exhibit three types of patterns, spatially sectoral or in the same sector. In the dry season they can be anastomosing because the sediment load exceeds the discharge, and in the wet season they can be meandering (CUNHA, 2001).

According to Christofoletti (1980), channel types are the formats in which channels standardise their spatial arrangement along the river from the source to the mouth.

Straight channels

According to Cunha (2001), natural rectilinear channels are rare and are represented by short channel segments, most of which are controlled by tectonic lines or lowland channels controlled by sandy ridges in the deltaic plains.

According to Cunha (2001), the occurrence of straight channels is generally associated with a homogeneous bedrock that offers the same resistance to the action of the water. In non-rocky straight channels, the splitting of the channel from one bank to the other forms a cross-sectional profile with areas of greater depth and shallower, more comfortable areas, so that the two sides of the channel alternate and sediment banks are formed, as shown in Figure 06.

Straight channels are those in which the river flows in a straight line and does not deviate significantly from its normal course in the direction of the estuary (CHRISTOFOLETTI, 1980).

Figure 06 - (A) Representative model of a straight channel, (B) straight channel in the natural environment based on satellite images.

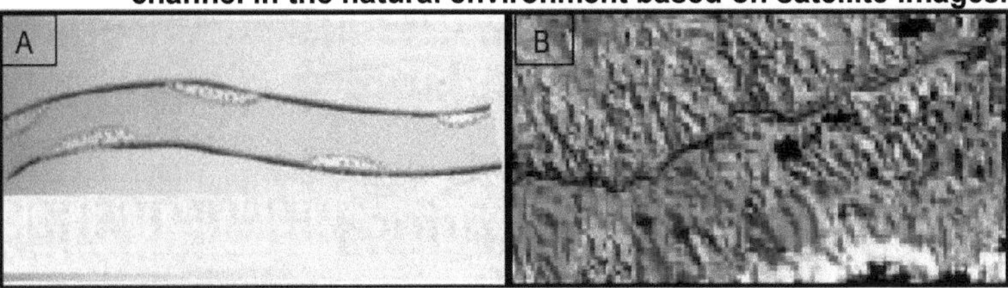

Source: Figure A, channel model based on Cunha (2001); Figure B, Landsat image.
Org.: PERETTO, A. (2011).

Meandering channels

These channels are often found in wetlands with riparian vegetation. They draw sinuous curves with harmony and similarities between them, with only one channel that overflows during floods. Along the riverbed, they have eroded and deposited banks that represent the first stage of the meander formation process. These meanders are formed where there are layers of sediments with mobile grain size, which are coherent, firm and not loose and have a low gradient with continuous and regular flow, with suspended and bottom sediments present in more or less equal quantities. When channels have these meandering forms, they are stabilised (CUNHA, 2001).

The development of the curvatures is uneven along the transverse sections, but in the straight sections between two continuous meanders the channels are asymmetrical, flat and there are riffles. In meanders, the profile is asymmetrical on the concave and deeper banks and becomes moderate towards the convex bank. In this type of channel, the transport of fine and selected sediments dominates, as the transport capacity is low in this section of the river and therefore a load of selected and uniform material is deposited (CUNHA, 2001).

For Christofoletti (1980), meandering channels are those in which the river describes sinuous, wide, harmonious and similar curves in its course, which is due to the continuous work of hollowing out the concave bank and depositing on the convex bank.

The meandering form, as shown in Figure 07, is more common in the lower reaches of the rivers, which is due to the fluvial behaviour and typology of the channels. In the upper and middle reaches, coarse sediments predominate and the channels are wider, shallower and have lower curvature indices. In the lower reaches, fine suspended sediments predominate and the channels are deeper and narrower. Due to these factors, they have higher sinuosity indices (CHRISTOFOLETTI, 1980).

Figure 07 - Meandering channels, sections of the Paraguay River in the area of the confluence with the Cabagal River, downstream of the urban area of the municipality of Cáceres-MT.

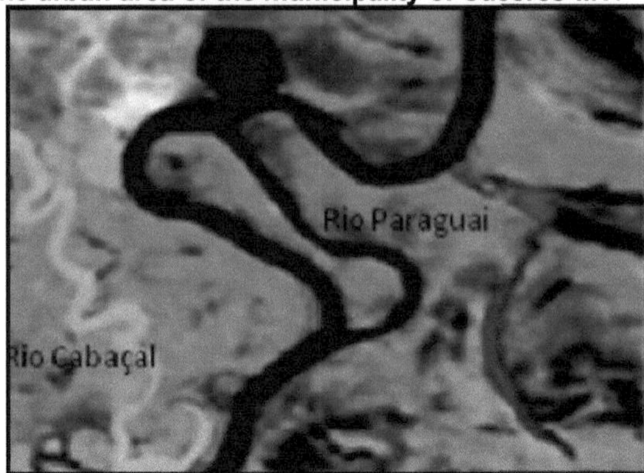

Source: Landsat satellite image. Org.: PERETTO, A. (2012).

In the work of Justiniano (2010), which aimed to determine the morphological changes that occurred over time in the Paraguay River between the mouth of the Sepotuba River and the Cabagal River from 1986 to 2008, he concluded that the floods of the Paraguay River cause periodic flooding, fluvial erosion and also an increase in sediment transport capacity. In the dry season, the sediment is deposited near the channel and on the plateaus. These are the factors that lead to changes in the channel, such as: a break in the meander neck, the occurrence of runoff, abandoned meanders, deposits near the active channel that form lateral accretion bars, edge dams and overflow deposits, and clogged meanders.

MORPHOMETRIC INDICES

Morphometric indices provide important data on river catchment areas. They help us to understand the characteristics of drainage channels, determine the capacity of the drainage network to discharge water and analyse the probability of flooding in a river catchment area.

Drainage density

According to Horton (1945, after CUNHA, 2001), runoff density is the index resulting from the total length of the channels in relation to the area of the catchment. This variable is directly related to the environmental characteristics of the area (climate, geology, geomorphology, soil type, vegetation cover and human activity). Where there is more impermeable rock, the conditions for rainwater runoff are better, allowing the formation of gullies and consequently an increase in runoff density. The opposite is the case with coarse-grained rock.

According to Christofoletti (1980), drainage density correlates the total length of drainage channels with the area of the river catchment and is an important index as it has an inverse relationship with the length of rivers. The higher the density of the rivers, the more the size of the fluvial components of the catchments decreases almost proportionally.

According to Horton (1945, op. cit., CUNHA, 2001), small catchments with a lower hierarchical level tend to have higher Dd values. According to

Christofoletti (1980), the lower order sections are located in the higher sectors of the catchment, which have a steeper gradient; as the area of the catchment increases and the hydrographic order increases, the channels cross areas of gentler topography, creating lakes with lower discharge density.

Collares (2000) carried out a study on the assessment of changes in micro-catchments as a tool for the geo-ecological zoning of river catchments: Application to the Capivari-SP river basin. The morphometric variables runoff density and flow density were used to evaluate the changes in the river basins and it was concluded that these two parameters can be expressed according to the composition of the subsoil. Micro-catchments with sandstone substrate had low values for runoff density, while micro-catchments with granite-gneiss substrate had higher values. The same behaviour can also be seen for flow density.

The density of the rivers

The purpose of obtaining information on river density is to compare the frequency or quantity of watercourses in an area. This index was first defined by Horton (1945, apud CHRISTOFOLETTI, 1980) and is also called hydrographic density by other authors. The density of rivers is related to geological conditions, geomorphological aspects, the availability of water (precipitation) and the degree of use/occupation.

This parameter relates the number of rivers or channels to the area of the catchment and expresses the size of the hydrographic network by indicating the capacity to generate new watercourses depending on the pedological, geological and climatic characteristics of the area (FREITAS, 1952).

For Christofoletti (1980), flow density is the variable that can reveal the hydrological behaviour of a catchment in relation to one of its fundamental factors, namely the willingness to create new drainage paths.

With regard to anthropogenic changes in a catchment, the temporal changes in Dr values will be reflected in the other morphometric variables.

Circularity index

This index was proposed by Müller (1953, apud CHRISTOFOLETTI, 1980) and represents the ratio between the total area of the catchment and the perimeter of the boundary, which corresponds to the total area of the catchment that is best related to the river in terms of areal extent.

Christofoletti (1980) also referred to the circularity index as the shape index and found that the closer the result found is to the unit Ic=1.0, the more circular the basin is.

The roundness index is simultaneously related to the compactness coefficient and tends to unity 1.0 when the basin approaches a circular shape and decreases as the shape becomes more elongated (CARDOSO et al., 2006).

This index is important for determining whether a particular catchment area is at risk of flooding or not, because depending on its geometric shape, it is possible to determine whether or not it is at risk of flooding from the overflow of the main channel.

Curvature index

According to Alves et al. (2003), the sinuosity index of channels is related to the influence of the sediment load and also to the lithological compartments, the arrangement of the geological structures and the slope of the channels. They also describe that values close to 1.0 indicate that the channel tends to be straight. In contrast, channels with values above 2.0 indicate that they are sinuous channels, and intermediate values indicate transitional forms between the regular and irregular types.

According to Villela and Mattos (1980), this index makes it possible to describe the degree of tortuosity of watercourses. The tortuosity of drainage channels is one of the factors that make it possible to control the speed at which the water flows through the basin.

Hierarchy of rivers

According to Christofoletti (1980), the river hierarchy is the process of categorising a particular watercourse within the entire river catchment in which it is located. This practice makes morphometric work on river catchments more objective.

For Cunha (2001), river planning should be the first step in conducting studies of morphometric analysis of river basins (areal, linear and hypsometric analysis), and he reports that this perspective of quantification was greatly expanded in the 1960s and 1970s, when many studies were conducted by researchers from Brazil and abroad; examples include: Freitas (1952), Schumum (1956), Strahler (1950, 1952b, 1956, 1957), Shereve (1956), Gandolfi (1971), Christofoletti (1969, 1970) and Cunha (1977, 1978).

According to Christofoletti (1980), it was Robert E. Hortan who defined and proposed the first criteria for a more precise categorisation of watercourses in 1945. According to this, first-order channels correspond to sources where the volume of water is still low and there are no tributaries; second-order channels correspond to the intersection of first-order channels and only accommodate channels of this order. Third-order channels are those that result from the intersection of two second-order channels and receive second- and first-order channels. Fourth-order channels result from the intersection of two third-order channels and receive third-order and lower-order channels. For Horton, each time a channel ascends, the entire length of the main channel must be considered, with the numbering being redefined from the last confluence of the channel to the source of the longest tributary.

However, some scientists felt that Horton's classification required subjective decisions, so they adopted Strahler's (1952) classification proposal, which suggested a classification in which the smallest channels without tributaries from the source to the confluence are considered first-order channels; second-order channels arise from the confluence of two first-order channels and receive first-order channels; third-order channels arise from the confluence of two second-order channels and receive second- and first-order channels. Fourth-order channels arise from the confluence of third-order rivers and receive lower-order tributaries, and so on in the higher orders that occur in the catchment area. With this classification, Strahler abolished the concept that the main river had to have the same number of orders along its entire length (CHRISTOFOLETTI, 1980).

GEOGRAPHIC INFORMATION SYSTEM

For some scientists, the definition of GIS can also be referred to as geoprocessing, as it is the same thing.

Geographical information systems (GIS) are automated systems that store, analyse and manipulate geographical data. These are data that represent phenomena and objects and that can be geographically located, whereby this location is the essential feature of the information and is indispensable for analyses (CAMARA et al., 1996).

Geoprocessing has become indispensable in the study of river basins, as it has methods that process images and facilitate interaction between the user and the system (COLLARES, 2000).

Geoprocessing is defined as a transdisciplinary technology because it integrates different disciplines, devices, programs, processes, units, data, methods and people who collect and process data and also analyse and present information in the context of geo-referenced digital maps by locating and processing geographic data (ROCHA, 2000).

According to Fonseca (2006), the use of GIS in environmental and water resources has become more widespread as they provide mechanisms to identify the spatial variability of river basin characteristics. Several studies have been carried out using river basins as the study area and using geoprocessing tools to obtain fruitful results.

LAND USE MAPPING

The IBGE (1999) defines a "land use survey" as the set of actions required to produce a thematic survey that can be synthesised through the production of maps. The land use and land cover survey shows how the land use typology is geographically distributed by identifying the homogeneous patterns of land cover, which includes both desk research and field research, the latter aiming to interpret, analyse and record the observations of the landscape, thus characterising the land use and land cover types in order to classify and locate them on thematic maps.

According to Collares (2000), satellite images are a satisfactory tool for the production of land use and settlement maps. Landsat-TM imagery is suitable for mapping land use and settlement at a regional scale, while satellite imagery with better spatial resolution, including spot imagery, is suitable for delineating surface features of the terrain, e.g. for surveying and delineating the drainage network of river catchments.

CHAPTER 4 THEORETICAL AND METHODOLOGICAL APPROACH

This work was developed through an integrated analysis of natural and non-natural elements based on geographical knowledge derived from geosystems theory.

All our reflections and actions and their realisation through our work are underpinned by the thoughts that are disseminated at the historical moment of the social group of those who carry them out. We achieve results that are underpinned by the theoretical and methodological concepts that are prevalent at the time of production.

Academic work must be developed through actions based on scientific knowledge. This knowledge is perfected over time with the scientific and intellectual maturity that each person acquires through the work they do. We must always look for better results, raise new questions and try to answer them. In doing so, we must always bear in mind the need to use the appropriate techniques and methods that are compatible with the objectives of the various topics under investigation.

THEORETICAL, METHODOLOGICAL AND TECHNICAL CONCEPTS

Geography as a science has always had the understanding of geographical space as its object of study. In the course of the historical development of geography, new paradigms and concepts have emerged that have gained in scope and importance. At the beginning of the 21st century, this science is trying to understand the dynamic changes in space and has included environmental issues as a category of analysis (SPIRONELLO and DE BIASI, 2005).

Today, countless changes are taking place in nature and in our globalised society that require more and more study to understand them, especially in the face of ongoing environmental disasters. These changes have a direct impact on humans because, according to the authors mentioned above, these changes can lead to the need to study and understand environmental issues. And to understand and conduct these studies on nature, more and more scientists are relying on geosystem theory.

After Spironello and De Biasi (2005), other followers emerged who excelled in the construction of geographical thinking; we can point out that in the middle of the 20th century: Bertrand (1971), Tricart (1977), Sotchava (1977), Christofoletti (1969) and Monteiro (2000).

According to Ross (2009), the geosystem was introduced in Brazil by Bertrand in 1971, through his work translated into Portuguese entitled: *Paisagem e geografia física global: esboço metodológico* (*Landscape and global physical geography: a methodological overview*), which was based on the construction of knowledge through "landscape science" and demonstrated that landscape studies should be based on the concept and methods of the geosystem. Bertrand introduced the concept of the geosystem to the Brazilian and French literature and built knowledge through 'landscape science' by trying to show that landscape studies should be based on the methodology of the geosystem.

According to Bertrand (1971), the theory of the geosystem is to be understood as "a clear, well-defined landscape that can be identified, for example, from aerial photographs" and as shown in the diagram in Figure 08.

The geosystem corresponds to the relatively stable ecological data resulting from the combination of geomorphological factors ("type of rocks and surface mantles, slope value, dynamics of slopes [...]"), climatic factors ("precipitation and temperatures [...]"), hydrological factors ("groundwater levels, springs, water pH, soil drying times [...]") and thus the ecological potential of the geosystem.

Figure 08 - Functional structure of the geosystem

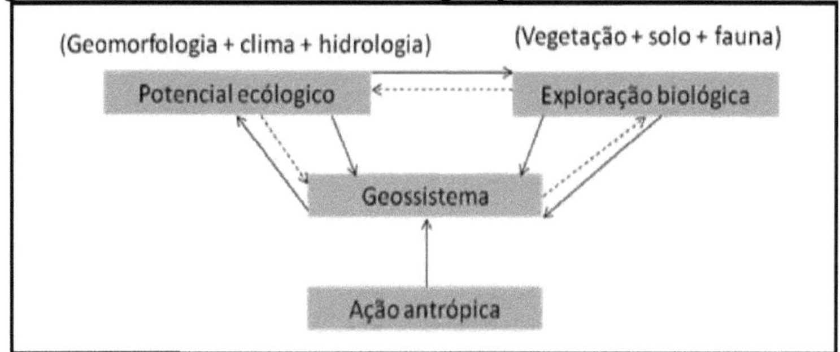

Source: BERTRAND (1971) apud ROSS (2009).

According to Bertrand (1971), studies based on the geosystem method involve analysing numerous variables resulting from the combination of geomorphology, climate, hydrology, soil, fauna and flora. The aim is to determine the ecological potential of a given area by establishing a balance between ecological potential and biological utilisation, which form an essential and dynamic unit both in time and space. In addition to these two factors, there is the human impact (ROSS, 2009).

According to Monteiro (2000), "the concept of geosystem is the integrating element in the geographical synthesis and also a factor in promoting interdisciplinarity in the understanding of the environment", but in order to work in an interdisciplinary way and gain an understanding of the different components of the environment, there are specific scientific fields to study these different components that make up the environment.

TECHNICAL PROCEDURES

The research was based on a series of integrated actions and, as a starting point, on the analysis of works for the theoretical and methodological basis of the work. This was followed by the selection of the research area and the analysis and study of its physical and natural characteristics.

According to Ross (2001), the methodology of a thesis represents the "backbone" of a research work, because in order to be able to apply a certain methodology, one must master the theoretical and conceptual content and also have the ability to deal with the instruments used in order to be able to distinguish between working techniques and methods.

OFFICE WORK

The cartographic basis was created using GIS (Geographic Information System). This software was used to create a database with the spot image MI 1924 from 2007 - this image served as the basis for the other products. Two software packages were used: Spring 5..7 and ArcGis 9.3.

The data on the drainage channels come from the SEPLAN-MT hydrographic database (year 2000), where corrections were made by superimposing hypsometric data that allowed a better vectorisation of the channels of the drainage network of Ribeirao Grande. The delimitation of the catchment area was carried out using the satellite image Spot, scene MI 1924, with a resolution of 2.5 metres from 2007. The extent of the catchment area and the length of the channels were determined using the Spring 5.1.7 software via the "Tool for metric operations" application, as shown in Figure 09.
Elaboration: PERETTO, A (2011).

Figure 09 - Manipulating the Spring software with the Operations tool Metrics

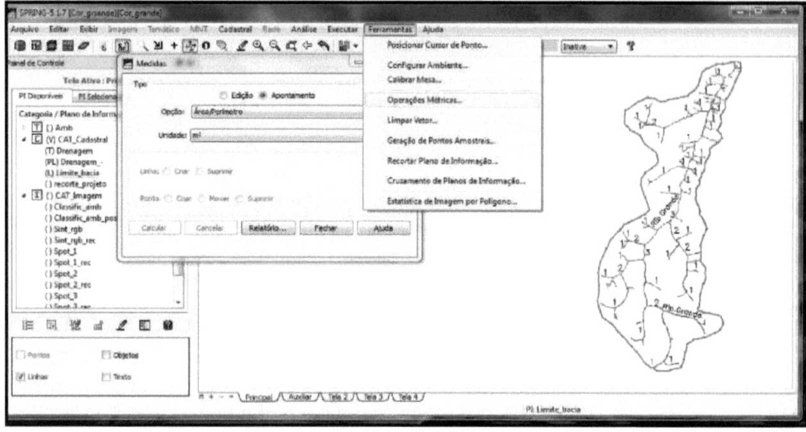

The calculation to determine the drainage density was established by Horton (1945) and can be calculated using the following equation:

$$Dd = \frac{Lt}{A}$$

Dd stands for the discharge density, *Lt for* the total length of the channels and *A* for the area of the catchment area.
[2]For the Dd classification of the river catchments, the parameters established by Christofoletti (1980) were used, i.e: Dd < 7.5 (km/km) - low; Dd 7.5-10 (km/km2) - medium and Dd > 10 (km/km2) - high.

The density of the rivers was calculated using the existing formula of Horton (1945), which establishes the relationship between the number of rivers and the area of the catchment area:

$$Dr = \frac{N}{A}$$

Dr is the density of the rivers, N the number of rivers or streams and *A* the area of the catchment under consideration.

The formula proposed by Müller (1953, apud CHRISTOFOLETTI, 1980) was used to determine the circularity index.

$$Ic = \frac{A}{Ac}$$

Where *A* is the total area of the catchment and *Ac* is the perimeter area of the catchment circle. This index shows the relationship between the total area of the catchment and the area of the catchment circle and is directly related to the river discharge.

According to Müller (1953), this index has a comparative parameter expressed as follows: (I) Ic = 0.51 - moderate discharge, with a low probability of rapid flooding; (II) Ic > 0.51 - circular basin with a high probability of flooding and (III) Ic < 0.51 - basin with a more elongated shape that favours water flow.

This parameter can be used to check whether the basins are shaped in such a way that flooding can occur. Ic = 0.51 therefore stands for a moderate discharge and does not contribute to a water concentration that allows flash floods. Values below 0.51 indicate that the catchment area is rather circular, which favours flooding processes (flash floods).

The sinuosity index establishes the relationship between the length of the main channel and the vectorial distance between the two extreme points of the main channel (SCHUMM, 1963, apud ALVES and CASTRO, 2003):

$$Is = \frac{L}{dv}$$

Where L is the length of the main channel, *dv* is the vector distance between the end points of the main channel.

The division of the catchment areas into sectors, upper, middle and lower reaches makes the spatial allocation and localisation easier to understand and enables the data on the catchment areas and the various topics to be categorised.

The sectorisation of the catchment was based on the hypsometric map mentioned above, and the sectors were divided according to the following altitudes: (I) upper reaches, above 380 m; (II) middle reaches, from 380 to 340 m; and (III) lower reaches, below 340 m.

The average channel valley width measurements were obtained by measuring the width of the channel valleys at 10 points in each sector of the basin. Then all points in each sector were added together and divided by the number of points to obtain the average channel valley measurement. To obtain this measurement, satellite images were used with an overlay of the MNT (numerical terrain model) image, which can be referred to as a radar image, and the measurement was performed using the Metric Operations tool in Spring 5.1.7, as shown in Figure 09.

The criteria proposed by Horton (1945) and modified by Strahler (1952, cited in CUNHA, 2001) were used to organise the hierarchy of the river network. For Strahler, the channel sections that form are those without tributaries that do not receive tributaries. They are rivers with a first-order hierarchy, and when two first-order channels converge, channels with a second-order hierarchy are formed that only accommodate first-order tributaries. If two second-order segments converge, a third-order channel is created that only accepts lower-order channels. If two channels of the same order converge, a channel of a higher hierarchy is created.

The topographical profile was created based on the elevation change of each sector of the basin: The vertical column shows the elevation change of the terrain in metres, the horizontal column shows the distance in kilometres over which the elevation change occurred. The profile was created using MNT data.

Arcgis 9.3 software was used to create the hypsometric map, in the ArcCatalog module, function: Spatial Analyst Tools - Surface - Contour. The software was instructed to create a digital file with contours at a distance of 20 metres. The map created in this way can be seen in Figure 22.

To analyse the geographical distribution of the main soil types in the Ribeirao Grande catchment, the work of Cabral and Cabral (2010) and Cabral and Silva (2011) was taken into account, which sought to understand the geographical distribution of the main soil types from the point of view of their correlation with the variations in the amplitudes of the interfluvial surfaces in the Parecis Plateau region, using the municipalities of Sorriso and Vera as examples.

With the availability of and easy access to satellite imagery, land use mapping using GIS (Geographic Information System) has quickly and efficiently provided data and information on land use and highlighted environmental problems such as deforestation of permanent protected areas, soil erosion, river siltation and others.

GIS information can also help to mitigate environmental problems and plan colonisation according to the potential of a particular area, and it can provide data for crop estimates, which is economically useful for concluding trade agreements.

The land use map was created on the basis of the spot image MI 1924 from 2007 using the Maxver classifier in the Spring 5.1.7 software. Three land use classes were defined that were easy to identify and classify: agricultural use, preserved vegetation and water. Each class has specific characteristics in terms of texture, shape and pattern, which facilitated the process of classifying and reclassifying the image. This process also involved checking or validating the data in the field by collecting points for validation, which were geo-referenced using GPS (Global Positioning System).

4.1 LABORATORY ACTIVITIES

The collected material samples were analysed using the pipetting method according to Embrapa (1996), which consists of determining the amount of sand, silt and clay fragments. To determine the result, 20 g of sediment is weighed in a 250 ml plastic beaker, placed in a beaker and 100 ml of distilled water with 10 ml of sodium hydroxide is added, then shaken with a rod and left overnight.

The next step was to pour the contents into the shaker, add 150 ml of distilled water and shake for 15 minutes. The shaken sample was then transferred to a 1000 ml beaker, sieved through the 270 mm sieve (equivalent to 0.053 mm), the material retained on the sieve was washed without exceeding the 1000 ml mark and the retained sand was placed in a pre-weighed beaker. The retained sand is then placed in the oven to dry. After the drying time has elapsed, the material is weighed and separated into coarse, medium and fine sand.

The material in the beaker is stirred by hand, monitoring the temperature during the waiting time for pipetting, and 25 ml is pipetted to a depth of 5 cm. The pipetted material is placed in a weighed vessel and left in the oven until the contents have evaporated; the vessel is then removed, cooled and weighed on a balance.

FIELD ACTIVITIES

The fieldwork in the Ribeirao Grande catchment area in the municipality of Sorriso-MT took place on 25 and 26 November 2011. In order to obtain an overview of the characteristics of the valleys of the drainage channels, morphometric surveys were carried out, bed sediments were collected, images were taken with georeferenced photos and GPS points, and data were collected through interviews. In addition, descriptions and analyses of the elements that make up the work were carried out, the classification of the land use map was validated and the main environmental problems were studied.

The interviews were conducted in informal conversations, without a predefined script for the questions asking about topics relevant to the research. The people who were interviewed were those who were encountered along the route of the field study.In general, the fieldwork on the first day focussed on the upper reaches of the basin. The route taken by car on 25 November 2011 was as follows: It began in Sorriso in the Primavera district, where the Delicious Fish farm was visited, and continued towards the town of Lucas do Rio Verde, where it turned right towards Vale do Verde 7.5 km after the Primavera district, It then continued towards BR 163, where it turned off next to the federal police station, and continued on BR 163 towards Lucas do Rio Verde, where it turned right again 4.3 km after the federal police station and continued to Ribeirao Grande, passing the Galheiros stream.On 26 November 2011, the route focused more on the northernmost part of the hydrographic unit in question. On that day, the route left Sorriso in the direction of the Teles Pires river, passing the toll booth, towards Ipiranga do Norte, to turn right after about 12 kilometres and cross the Ribeiráo Grande. Finally, we head north-south along a longitudinal route through the basin, crossing the lower and middle reaches and exiting at the federal police station next to the BR 163 motorway. The route travelled by the field class is shown in Figure 10 below. It shows the routes that led through the upper, middle and lower reaches of the basin on 25 and 26 November 2011.

Figura 10   - **Map of the field route recorded on 25 and 26 November 2012 in the Ribeirao Grande catchment area**

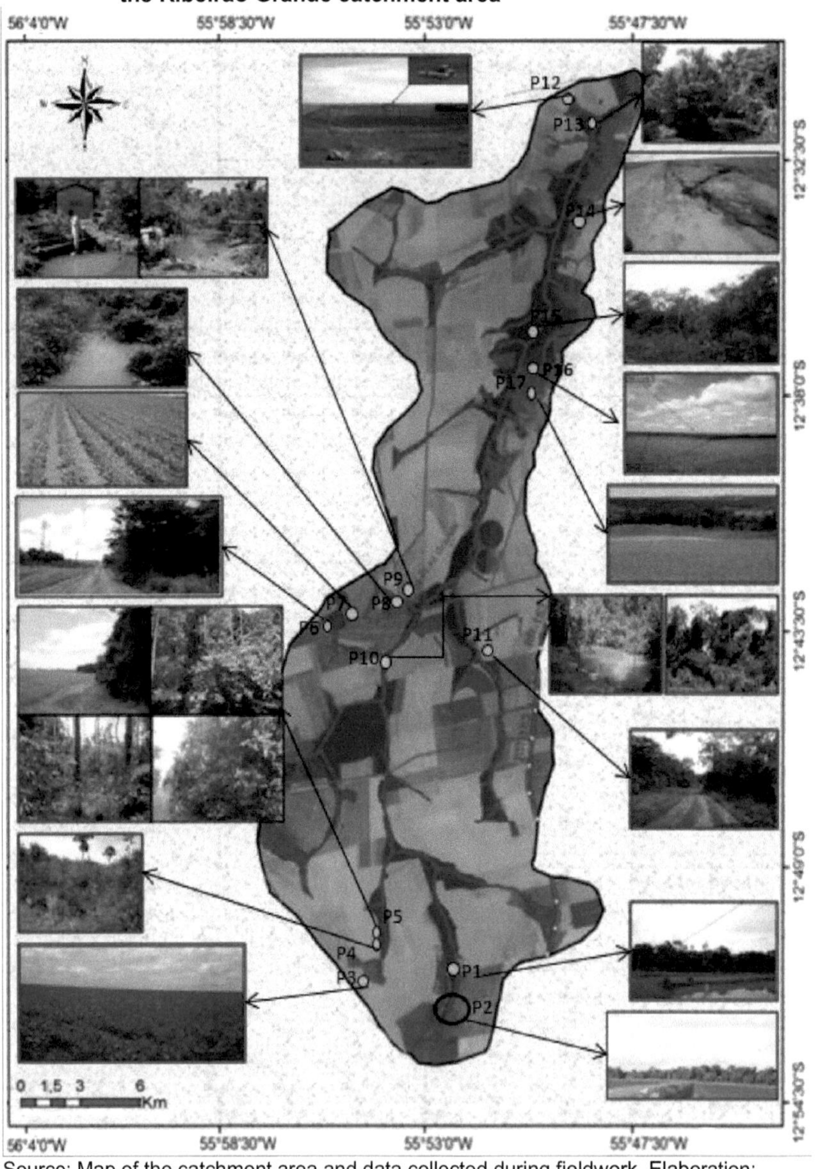

Source: Map of the catchment area and data collected during fieldwork. Elaboration: PERETTO, A. (2011).

CHAPTER 5 RESULTS AND DISCUSSION

The Ribeirão Grande catchment was the subject of this study. In this section, tables, figures, maps, diagrams and profiles based on data from field research, literature and laboratory data are presented to support the explanations and discussions of the results of this work.

5.1 DATA AND MORPHOMETRIC INDICES OF THE RIBEIRAO GRANDE CATCHMENT AREA

In order to understand the reality of the morphometric indices of the Ribeiráo Grande catchment, a survey of the parameters required to calculate the indices was carried out. Table 01 shows the results of the data found.

Table 01 - Measurement of morphometric data of the Ribeiráo Grande catchment in Sorriso-MT

MORPHOMETRIC DATA	RESULTS
Circumference	119.2 kilometres
Bay area	415.3 kilometres^2
Total length of the pipe	208.9 kilometres
Number of canoes	72
Length of the main channel	51 kilometres
Vector length of the main channel	42 kilometres

Source: Based on the information in the database created from the satellite image SPOT 5 (2007).
Elaboration: PERETTO, A. (2011).

The perimeter of the border circle of the Ribeirao Grande basin is 119.2 km long. [2]The area of the basin is 415.3 km, which shows that it is not a very large area.

The total length of the 72 channels in the Ribeirao Grande catchment is 208.9 km and can be considered a catchment with a small total length of channels. This value was one of the data points used to calculate the drainage density of the basin.

The main channel of the Ribeirao Grande runs from its source about 4.6 km west of the Primavera district, near the Delicious Fish farm, for 52 km to its mouth, where it flows into the Teles Pires river on the left. The vector length, i.e. the distance in a straight line from the main source to the estuary, is 42 km.

The morphometric indices calculated in the Ribeirao Grande catchment were used to analyse the most important morphometric indices; their values are listed in Table 02.

Table 02 - Results of the morphometric parameters of the Ribeirao Grande catchment

MORPHOMETRIC INDICES	RESULTS
Drainage density	0.50 km/km2
The density of the rivers	0.17 channels/km2
Circularity index	3,4
Curvature index	1,4
Sequence of channels	[a]4 Order

Source: Based on the information in the database created from the satellite image SPOT 5 (2007).
Elaboration: PERETTO, A. (2011).

[2]The Ribeirao Grande catchment has a discharge density of 0.50 km/km, which indicates a low discharge density. [2]It should be noted that, according to Christofoletti (1969), river catchments with a Dd < 7.5 (km/km) should be considered as having a low discharge density.

The low runoff density of the Ribeirao Grande basin is related to Christofoletti's (1980) explanation that the hydrological behaviour of the rocks in the same climatic environment affects the low runoff density. Thus, in areas where rainwater infiltration is more difficult, favourable conditions are created for surface runoff, leading to the formation of gullies - an example of this is fine-grained clastic rock.

In the Ribeirao Grande catchment area, on the other hand, the sediments of the detrital laterite cover layer of the Parecis Group, which have a medium to coarse unconsolidated granulometry. This type of rock facilitates the infiltration of water and impedes the excavation and formation of new channels, resulting in a low runoff density in the study area.

[2]The low discharge density of 0.50 km/km of the Ribeirao Grande is also related to its location in the Teles Pires catchment, as in the macro context of its location it lies in the middle course, a course in which the number of channels and thus the river and discharge density decreases.

This is due to other factors, such as structural factors (the sedimentary structures of the Parecis group favour infiltration overall) and the number of streams, which is also relatively low. In the middle reaches of the catchments, the length of the channels is shorter and consequently the discharge density (Dd) and the density of the rivers (Dr) are lower.

[2]The catchment area of the Ribeirao Grande has a flow density of 0.17 channels/km - this figure shows that it is a catchment area with a low flow density index.

The relatively low flow density of the Ribeirao Grande catchment shows that the ability to form new channels in this catchment is very low. This factor is related to the geomorphology of the area, which has a predominantly flat, tabular relief that makes the formation of new channels difficult. The catchment area has a high infiltration capacity due to its lithological composition, which consists of sediments with a coarse, unconsolidated granulometric texture from the detrital-lateritic formation.

The circularity index value in the Ribeirao Grande basin was 3.4, which is a high value and shows that this basin has an elongated shape.

The circularity index is intended to show whether flooding is possible in a particular catchment area. If the result is greater than 0.51, this means that the area has only a low probability of flooding. Since the value of this index obtained in Ribeirao Grande is well above the value established as a parameter, it can be concluded that the catchment, which has an elongated shape, has no possibility of flooding in this area, and this factor is related to the configuration of the relief shapes, which are flat and have a low slope index, which is parallel to the main axis of the macro-discharge of the Teles Pires River in the S/N direction.

The valleys of the channels in the Ribeirao Grande catchment are wide, and in order to better understand them, the average width of the valleys in each sector of the catchment was measured. According to the data in Table 03, the average width of the valleys in the general context of the basin is over 100 metres. In the upper, middle and lower reaches, the valleys are 111, 172 and 259 metres respectively.

Table 03 - Width of the valleys by sector of the lower, middle and upper reaches of the Ribeirao Grande

Sectors of the catchment area	Average valley value (m)
High exchange rate	111
Medium course	172
Low stroke	259

Source: Based on the information in the database created from the satellite image SPOT 5 (2007).
Elaboration: PERETTO, A. (2011).

The division of the Ribeirao Grande catchment area into high, medium and low sectors is shown in Figure 11, whereby this division was made on the basis of the hypsometric parameters of the terrain.

The upper reaches of the Ribeirao Grande basin form the largest land area, which is over 380 metres above sea level. Most of the springs and very wide catchment areas are located in this area. The middle course lies in the hypsometric classes between 380 and 340 metres above sea level and occupies the second largest area of the basin. The lower reaches occupy the smallest area of the basin, with altitudes below 340 metres. This is where the Ribeiráo Grande flows into the Teles Pires.

Figura 11 - **Subdivision of the Ribeirao Grande catchment area into sectors: upper, middle and lower reaches**

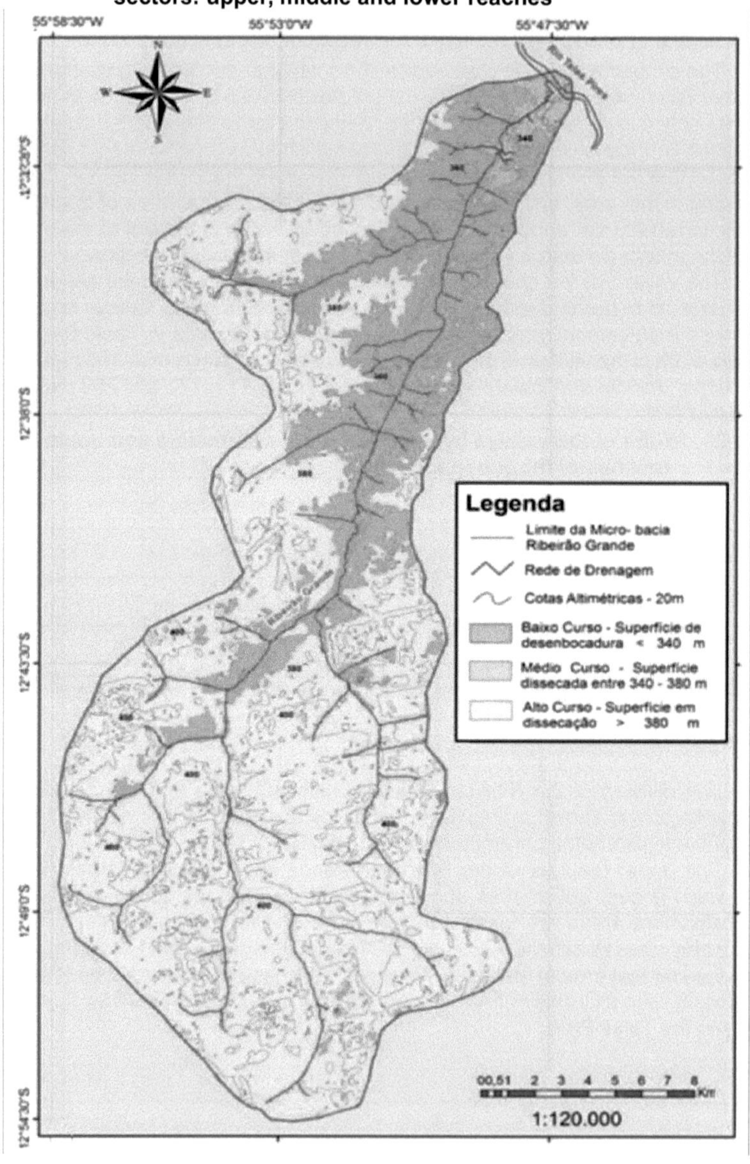

Source: Data based on an MNT image (numerical terrain model). Elaboration: CABRAL, T. L. (2011).

In the upper part of the Ribeirao Grande basin are the smaller valleys with an average width of 111 metres, where the dynamics of relief formation by drainage through the process of degradation/incision predominate.

The valley incision system in the various sectors of the catchment area, i.e. on the surfaces of the springs - the headwater sector - is dominated by surface degradation processes. This surface degradation occurs where a large part of the valley incision system is in full vertical action.

In the middle reaches of the river there is a certain balance between what happens in the upper and lower reaches, and the valleys of the canals have an average valley width of 172 metres.

In the lower reaches, the process of action of the hydrographic system is a pleasant one, i.e. the rivers, or rather the main river, begin to play the role of surface build-up, resulting in a series of indicators of these events, including the configuration of the valleys, which become wider.

The average survey of the valleys is important because it helps to get an idea of what the valleys of the canals in the Ribeirao Grande catchment area look like, making it possible to discuss how important this was for the conservation of the riparian vegetation of the canals.

The analysis of the satellite image of the study area and a site visit showed that more than 95% of the protected areas of the Ribeirao Grande basin are located in the valleys of the canals. This shows that these valleys have directly contributed to the fact that 19.5% of the basin's area is covered by preserved vegetation.

The drainage channels of the rivers in the Ribeirao Grande catchment area are located in wide valleys with an average width of over 111 metres and have sufficient surface area to drain the water, that may be surplus in the drainage channels of the rivers in this catchment at 12° 44' 16.1" south latitude and 55° 51' 43.5" west longitude, as can be seen in Figure 12, which shows the valley of the Córrego Galheiros, a tributary of the right bank of the Ribeirao Grande.

Figura 12 - **Valley of a source of the Córrego Galheiro (first order river). The section shows the configuration of the valley with a wider, flatter and more open aspect.**

Photo: PERETTO, A. (25/11/2011).

The valley shown in Figure 12 is a wide valley with well-preserved vegetation. In most of the valleys of the drainage canals in this catchment area, the vegetation has been preserved and has only been removed or killed by eutrophication where dams or road embankments have been built to cut off the canal valleys.

The natural vegetation of the drainage channels is preserved, a factor that contributes to the conservation of the channels. The vegetation along the watercourses reduces the impact of agricultural use in terms of the formation of sedimentation sources, sediments that would lead to the silting up of the channels, which would be one of the main problems favouring the occurrence of floods.

In the Ribeirao Grande catchment area, however, rainwater leads to flooding at the top of very wide catchment areas, as the gradient of the terrain for surface runoff is low. During the rainy season, the water often accumulates on the interfluvial areas, which have large and very large dimensions. The flooding differs from the conventional system of a flood inundating the main course.

The extraction of water from this part of the relief due to the topographical configuration with a low gradient between the source and the mouth, which corresponds to the type of morphology resulting from sedimentary structures on a passive platform in humid tropical environmental conditions, means that the problems associated with the surfaces that have the potential to accumulate water also become those of the upstream surfaces. To summarise, rainwater entering the dissected parecis plateau sector, and therefore the watershed in question, follows two paths in relation to the factors already mentioned: (I) accumulation on the surfaces of wide and very wide catchments and infiltration; and (II) runoff of water into the normal surface drainage system.

This type of flooding problem at the head of confluences also occurs in protected areas with natural vegetation, suggesting that flooding is not related to land use but to the design of the relief.

The flooded areas in wide incisions that occur during the rainy season make it difficult for plants to grow. This can be seen on satellite images from the rainy season, such as the Cbers image in Figure 13, where the flooded areas appear in dark blue tones, while the ploughed areas in good condition have light colour tones in relation to the water bodies.

Figura 13 **- Aerial view of the upper reaches of the catchment area, the floodplains and the Primavera-MT district**

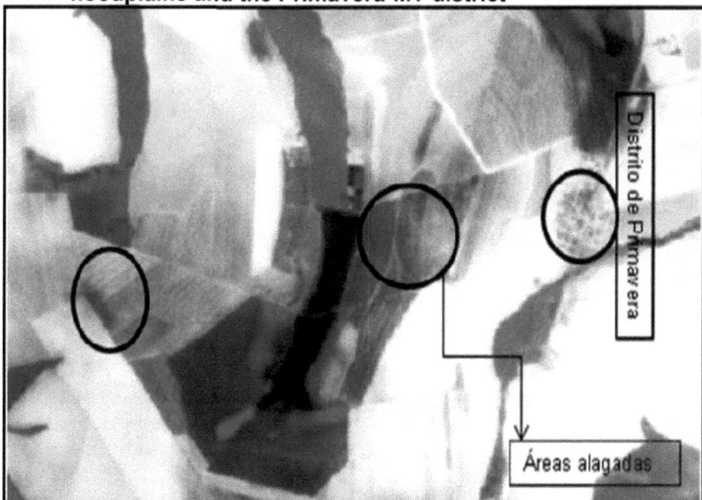

Source: Cbers image CCD1XS_16/09/2008_166_114.
Elaboration: PERETTO, A. (2012).

The presence of these floodplains shown in Figure 13 can be verified by cross-referencing the data from the 2.5 resolution spot image in Figure 14, which shows the presence of numerous artificial drains in the same area of the upper Ribeirao Grande catchment. The artificial drains in the image are the straight lines highlighted in the circle, as if they were roads, but they are artificial drains created by farmers and located in ploughed areas.

Figura 14 - **Aerial view of the upper reaches of the basin with drains, fish ponds and the Vera-MT district**

Source: Imagem Spot 2007.
Elaboration: PERETTO, A. (2012).

Artificial drainage channels are man-made channels with a width of around 2 metres and a depth of 1 to 1.60 metres. They have a rectilinear shape and are connected to drainage channels. Artificial drainage channels are created in the shallow areas of wide and very wide catchment areas. Their purpose is to drain rainwater that does not run off superficially and is not quickly absorbed by groundwater, as shown in Figure 15, making it impossible to grow soya, maize and other crops that grow in areas with well-drained soils.

With this technique of intervening in the environment through the construction of artificial drainage systems, man has enabled the cultivation of soya and maize, which do not thrive in areas with highly waterlogged soils. However, this technique of interfering with the natural dynamics of rainwater in the Ribeirao Grande catchment could potentially have an impact on groundwater recharge in the region. No study has yet been carried out on the potential damage caused by this practice, but it could potentially alter the natural water dynamics in the region.

This practice occurs in all flat areas of the state of Mato Grosso.
Figure 15 - Clean runoff with little vegetation

Source: CABRAL, T. L. (February 2011). Location: Municipality of Sorriso-MT.

Cabral (2011) reports that the purpose of natural surface drainage can be to improve water flow through works that can be divided into two different classes: (a) the systematisation of the soil and (b) the construction of a drainage network. These works complement each other and have different characteristics depending on whether the area is irrigated or not. In the case of flat or hilly terrain, both are surface drainage, i.e. drainage systems designed to channel rainwater and drain areas with excessive moisture problems, which justifies the use of drainage systems in the southern part of the municipality of Sorriso.The springs in the Ribeirao Grande basin are typical of the flat areas of the Cerrado with wide valleys and diffuse springs, as can be seen in Figure 16, located at the coordinates 12° 51'05.1" south latitude and 55° 54' 06.4" west longitude. The diffuse springs do not have a defined location where the water flows out of the spring, which makes it difficult to characterise the spring areas for the conservation of PPAs (Permanent Preservation Areas).

Figure 16 - Diffuse headwaters of the Ribeirao Grande, located on the Favareto farm

Source: PERETTO, A. (25/11/2011).

In the Cerrado region in particular, there is great pressure to utilise as much flat land as possible for agricultural plantations. Most people think of a spring as the place where the water starts to flow, without realising that in the case of diffuse springs, there is an entire wetland that is also called a spring.

5.1.1 Flow hierarchy

The hierarchy of channels in the drainage network was based on the classification of Strahler (1957), with a modification according to Collares (2000), in which intermittent channels were taken into account, as Strahler only considered permanent flows in his study. All channels recognisable on satellite images that allow a linear water flow were considered as drainage channels. The inclusion of intermittent channels is due to the fact that they contribute to water flow in the catchment and it is these channels that are most susceptible to change.

The drainage network of Ribeirao Grande consists of 72 canals, both intermittent and perennial. [a]Among these channels, according to Strahler's classification (1957), there are 55 1st order channels without tributaries, 14 2nd order channels corresponding to the channels that receive 1st order tributaries and only 2 3rd order channels. It was found that the Ribeirao Grande can be categorised as a 4th order river.

In the hierarchical division of the Ribeirao Grande drainage network into first, second and third order channels, there is a greater number of first order channels. This factor may be related to the characteristics of the flat terrain, which makes it difficult to form new channels, thus maintaining the large number of first-order channels.

When analysing the characteristics of the drainage network of the Ribeirao Grande basin, it can be seen that the direction of the drainage system is south/north and its tributaries in the lower and middle reaches are arranged in an east/west direction. However, the Galheiros stream is located in the upper reaches of the basin and its drainage network is organised in the same way as the drainage network of the middle and lower reaches, as can be seen in the map of the hierarchical classification of the drainage network (Figure 17).

Figura 17 - **Hierarchical categorisation of channels according to Strahler (1954)**

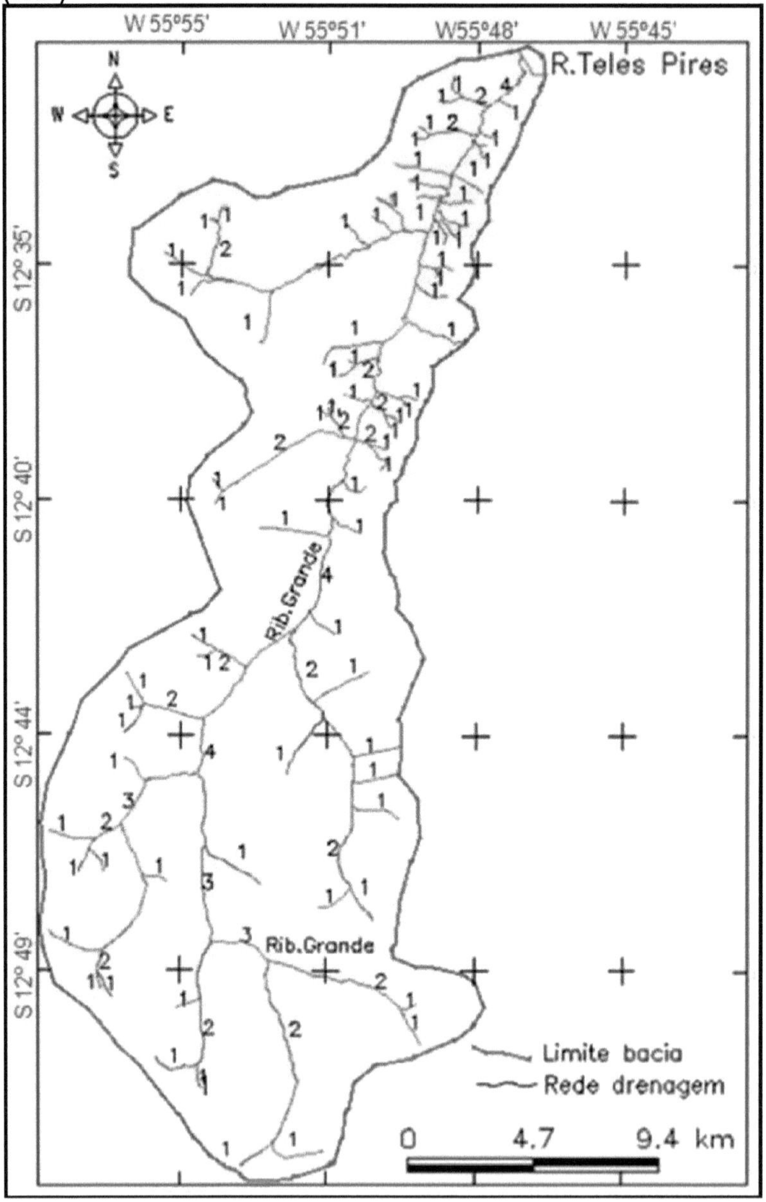

Source: Data from the Seplan-MT drainage network, corrected with the Spot MI 1924 image.
Adaptation and elaboration: PERETTO, A. (2011).

In the upper part, the drainage system and its tributaries run parallel to the Ribeirao Grande channel, along the major drainage axis that forms the Teles Pires river. This proves that the "spontaneity" of the geographical distribution of the drainage systems on the dissected Parecis plateau is controlled by a macro-regional structure.

[a]The problems caused by the construction of roads are evident in 1st order channels. In these channels, the river channel is not well defined or very shallow, which makes it easy for the water to flow laminar, and when the water flow increases during the flood period, the pipes installed in the roads have a low flow capacity and are only in the small channel, and when the water rises, it accumulates and floods the areas of the valleys that are wide, creating small dams along the roads, it accumulates and floods the areas in the valleys that are wide, creating small dams along the roads, a factor that may increase the frequency of flooding areas, not only in the areas of this basin, but also in other corresponding situations during the rainy season in the region.

The dam interrupts the natural flow of water and, especially during the flood season, this damming leads to the death of the riparian vegetation, as it is not adapted to the increased moisture conditions caused by the construction of the roads that separate the valleys, as shown in Figure 18 at coordinates 12° 51'05.1" south latitude and 55° 54' 06.4" west longitude.

Figura 18 [a]- View from the top of the embankment to the dying trees upstream after the construction of the road at the 1st order spring in Ribeirao Grande

Source: PERETTO, A. (25/11/2011).

The fact that road embankments "interrupt" the valleys of drainage channels is one of the effects that exist in the catchment area. These areas are the catchment areas of the higher parts where the roads are located. It is the secondary roads that have been cut into many of the valleys of the Ribeirao Grande springs, together with the very flat topography, that impede the flow of water in the upper areas of the Ribeirao Grande basin.

5.1.2 Drainage patterns in the Ribeirao Grande catchment area

Studies that investigate flow patterns lead to the identification of channel types and characterise the shapes that rivers take along their longitudinal course through the catchment.

This geometry of the channels shows the result of the work done to adapt the channels to their cross-section. These shapes reflect the result of various factors, such as: the geology of the area, the geomorphology and the type of climate and, in some cases, the impact of man on the drainage network.

The gradient index indicates whether the channels are straight or sinuous. In the Ribeirao Grande catchment area, the value of this index was 1.4, which indicates that the channels have intermediate values that indicate transitional forms between straight and sinuous types.

Straight channels predominate in the Ribeirao Grande catchment. According to Cunha (2001), natural straight channels are very rare and are represented by short channel sections, and in the Ribeirao Grande catchment the straight channels are first-order channels that are not very long, as shown in Figure 19 - Types of channels in the Ribeirao Grande catchment area

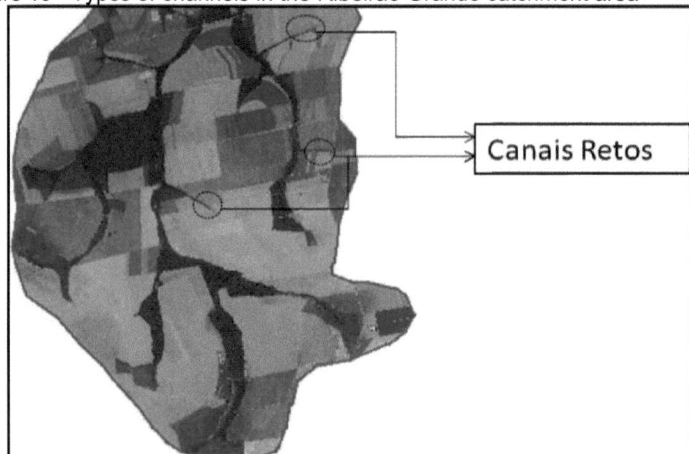

Source: Spot image scene MI 1924 from 2007, upper basin of Ribeirao Grande. Elaboration: PERETTO, A. (2011).

The main channel of the Ribeirao Grande alternates between straight and meandering sections. This type of straight channel is possibly related to the conditions of the relief, which is tabular and flat, and also to the fact that the rivers carry little sediment and suspended matter, so that sediment banks do not form in the drainage channels, a fact that may provide favourable conditions for the existence of this type of channel. The condition that favours the persistence of the drainage channels without major pollution by soil and suspended matter is related to the fact that the riparian forest has been preserved throughout the study

area.
PHYSICAL PARAMETERS OF THE RIBEIRÁO GRANDE CATCHMENT AREA
The Ribeirao Grande catchment area has physical relief features resulting from the geomorphological forms of the dissected Parecis plateau.

The gradient of the Ribeirao Grande catchment area is typical of regions with dissected relief that have a low gradient, such as the basins of the dissected Parecis plateau. In the study area, the gradient is 80 metres over a distance of 42 kilometres from the highest point of the springs to the confluence with the Teles Pires River, as shown in Figures 20 and 21.

The topographical profile in Figure 20 shows the differences in altitude in each sector of the basin: the vertical column shows the differences in altitude of the terrain in metres and the horizontal column shows the distance in kilometres over which the differences in altitude occurred.

Figure 20 - Topographical profile of the Ribeirão Grande basin

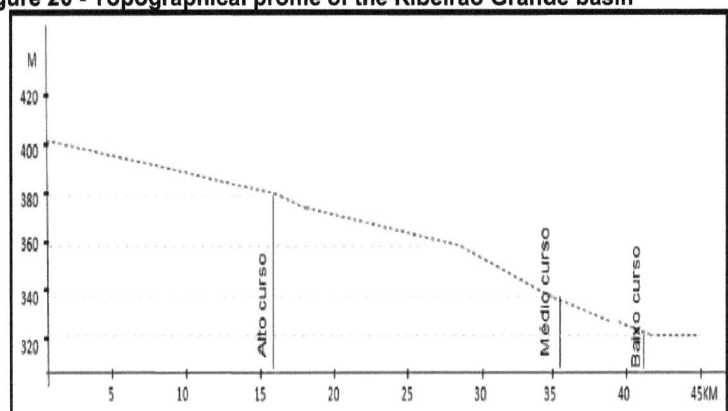

Source: Data from MNT images (numerical terrain model).
Elaboration: PERETTO, A. (2012).

In geomorphological terms, the group of forms present in the area in question refers to the system that forms the topographic units of the tabular and flat group of forms of the Parecis Dissected Plateau, as shown in Figure 21.

Figure 21 - Hypsometric map of the Ribeirao Grande basin
Source: Data based on an MNT image (numerical terrain model). Elaboration:
CABRAL, T. L. (2011).

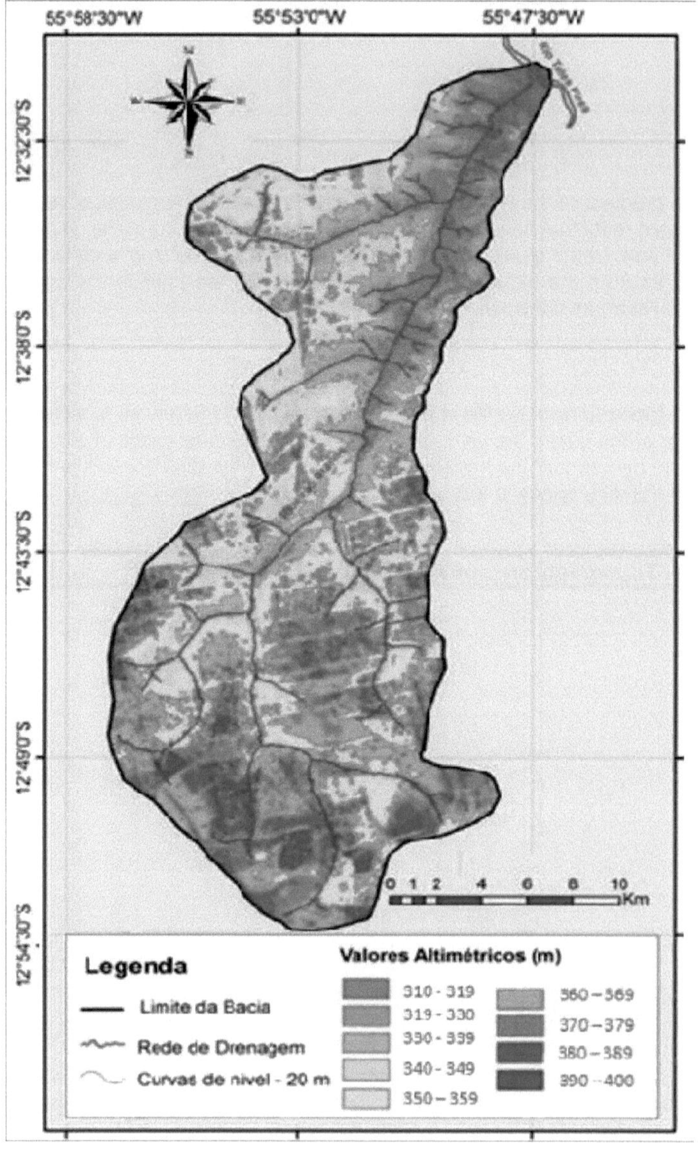

The sediment load transported by the Ribeirao Grande consists mainly of medium and fine-grained sand, silt and clay without organic matter. This can be seen from the data presented in Table 04, which were obtained by analysing bottom sediments collected in the lower part of the basin, in the main channel of the Ribeirao Grande, where a sample was taken at the coordinates 12° 32' 29.3" south latitude and 55° 48' 38.2" west longitude.

Table 04 - Results of the bottom sediment analysis for the Ribeirao Grande catchment area

			20g		
Sample	Coarse sand	Medium sand	Fine sand	Silt	Clay
1		14,34g	5,28g	0,04g	0,34g
2		13,76g	4,89g	1,25g	0,1 g
3		14,54g	5,19g	0,19g	0,08g

Source: PERETTO, A. (2011).

Coarse sand was not considered in this analysis, as there are no surface or subsurface structures with this grain size in the Ribeirao Grande catchment and there are no advanced erosion processes to remove coarser material.

The well-preserved condition of the riparian forests makes it difficult for the constituents to enter the drainage channels via the surface water runoff. The materials present are medium sand (75 %) and fine sand (almost 26 %), which originate from the natural erosion process of the canal banks by the watercourses.

As already described elsewhere, the catchment area of the Ribeirao Grande is characterised by well-preserved riparian and gallery vegetation, which is due to the fact that the springs only originate in the valleys, as can be seen in Figure 22, which is located at the coordinates 12° 37' 20.2" south latitude and 55° 49' 37.4" west longitude. In this close-up of the local landscape, two springs can be seen in the foreground and the Ribeirao Grande in the background, which drains its waters from left to right. In the close-up, a panoramic view can be seen in which the shapes of the tableland suitable for soya cultivation can be recognised, ideally reflecting the characteristics of the study area.

Figura 22 - **Panoramic view of the landscape in the catchment area of the Ribeiráo Grande, with flat and tabular landforms**

Source: PERETTO, A. (26/11/2011).

When analysing the Ribeiráo Grande basin, we note that in this landscape context we find the red/yellow and yellow latosol areas, which correspond to areas that are highly exploited for highly mechanised precision agriculture.

The Ribeiráo Grande basin has the following soil types: Red latosols, red/yellow latosols, yellow latosols and hydromorphic soils (glei and fluvial neosols).

The red latosols are located in the higher areas and in small and medium interfluves, which according to Cabral and Cabral (2010) have an interfluve extent of 2000 to 4000 metres. The red/yellow latosols occur together with the red and yellow latosol classes, which are associated with the topographic conditions of the relief at each site. Yellow latosols are found in areas with large and very large interfluvial extents that exceed 6000 metres.

The geographical distribution of latosol varieties is directly related to the variation of interfluvial amplitude in shallow areas and to regional climatic aspects. These factors are primarily part of an overall context that has made it possible to create the favourable conditions for the formation of latosol varieties in the morphostructural unit in which the Ribeirao Grande basin is located.

These factors have enabled the development of a state-of-the-art agricultural sector, making the municipality the largest soya producer in Brazil and the world. This strong production has brought a significant amount of resources to Sorriso, as soya is one of its export products. This factor has boosted the excellent production of the municipality's rural area, which is currently experiencing rapid urban growth and, consequently, the growth of sectors related to agriculture.

As described, the latosols offer good conditions for agricultural production. Figure 23, which is located at the coordinates 12° 43' 36.1" south latitude and 55° 55' 21.6" west longitude, shows a soya plantation in a medium stage of development in an area with wide and very wide transitions to the Yellow Latosol, which is located in the upper reaches of the Ribeirao Grande catchment area.

Figura 23 - **Yellow Latosol with a soya culture in the middle stage of development**

Source: PERETTO, A. (25/11/2011).

MAPPING AND CHARACTERISATION OF LAND USE TYPES

In river basin studies, land use surveys help us to understand how the land is used.

The data on proportionality in relation to the geographical distribution of the individual land use classes in the Ribeirao Grande catchment area are shown in the graph in Figure 24. According to the graph, the land use classes are distributed in terms of area as follows: The most extensive agricultural use accounts for 79% of the total area; the preserved vegetation accounts for 20% of the area; and the water class occupies 0.45% of the total area.

Figure 24 - Graph of land use classes in the Ribeirao Grande catchment area

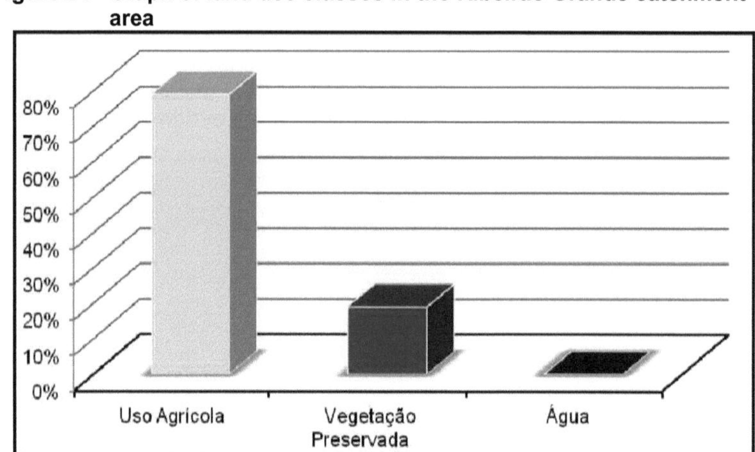

Source: Land use map of the Ribeirao Grande catchment area (2007).
Elaboration: PERETTO, A. (2011).

[2]According to the survey carried out, the largest type of land use in terms of area is agricultural use, with 330.6 km, representing 79% of the area of the Ribeirao Grande catchment. In this class, mechanised agriculture is predominant, with the use of high technology through the intensive use of machinery for the cultivation of crops in the catchment area.Another fact related to the means of production in the study area and in the region that should be highlighted is the use of irrigation, which has developed in the region as a viable alternative to increase the production of agricultural crops grown according to the rainfall calendar. With this method, producers implement a system of intensive production dynamics that makes it possible to include one or two more harvests a year in the agricultural production routine.

The class of preserved vegetation, which consists of green areas, has an area of 82.7 square kilometres, representing 19.5% of the total area of the catchment. It is also the class in which the APPs (Permanent Preservation Areas) are located; and the water class has a share of 0.45 % of the total area, as shown on the map in Figure 25.

Figure 25 - Map of land use classes in the Ribeirao Grande catchment area

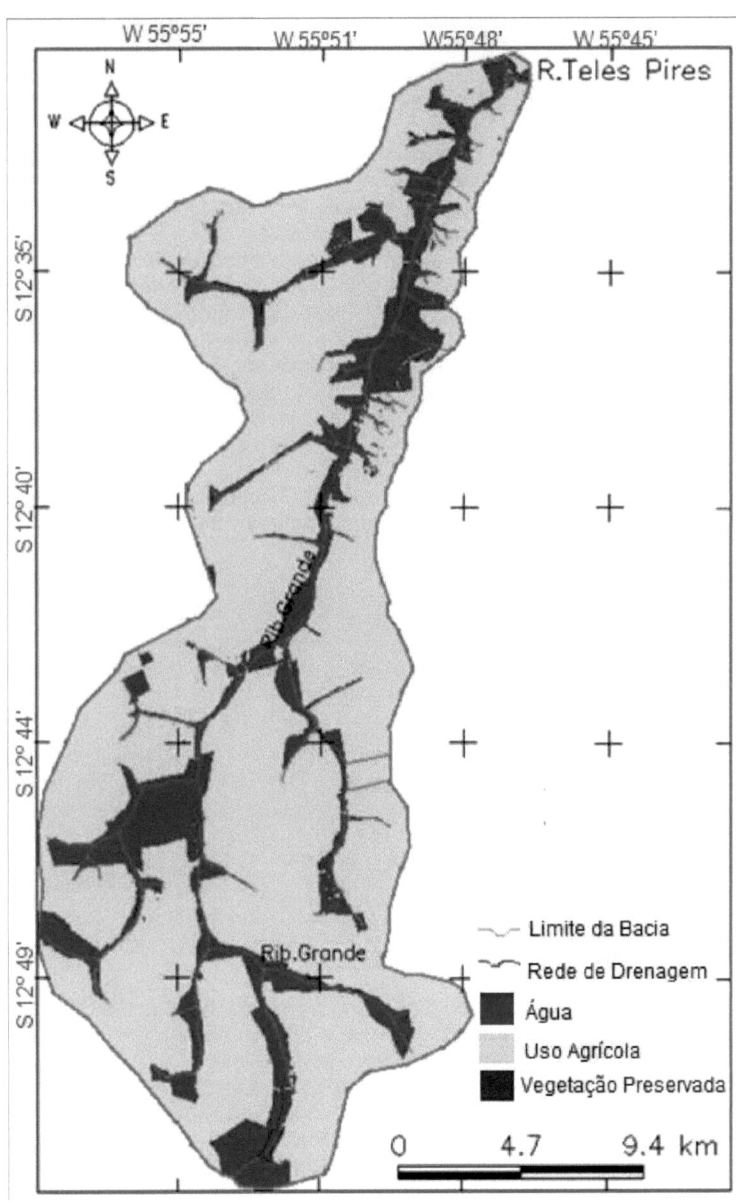

Source: Spot image MI 1924 from
2007. elaboration: PERETTO, A.
(2011).

The geographical distribution of vegetation cover in the Ribeirao Grande catchment is concentrated almost entirely, i.e. around 95%, in the valleys of the drainage channels, with only one area outside the drainage valleys in the lower reaches of the entire catchment studied being permanently preserved. This is due to the lack of favourable conditions for mechanisation in the valleys, a factor that has contributed to the preservation of riverbanks in all areas of the catchment, so that 19.5% of the total area of the catchment is covered by native vegetation.

The reason for the difficulties in using the land in the canal valleys is that these are very wet areas and that the canal is undefined, which means that the soil is very saturated in the rainy season, making it impossible to mechanise these areas in the valleys where the vegetation is preserved.

In the protected area of the lower reaches, this conservation is linked to the fact that the relief is very steep and also to the high concentration of iron dioxide outcrops (petroplintites), which make the area unprofitable for mechanisation.

Fish farming is another important economic sector, with the company "Delicious Fish", which began operations in 2004, playing a pioneering role. It currently produces fish on a large scale and has several ponds and dams with a water surface area of around 170 hectares (see Figure 26).

Figura 26 - **Fish breeding dams of the company "Delicious Fish", located in the district of Primavera, municipality of Sorriso-MT**

Source: "Delicious fish" (2011).

In terms of production, according to the managing director, the company sells around 4,000 tonnes of tambacú, pintado and tambatinga fish each year and is currently breeding pirarucu in an experimental phase.

At the fish farm in the Primaverinha district, soya and maize are grown on the land around the dam and the fish farming ponds. Everyone knows that these crops require intensive use of agrochemicals that are applied from the air by aeroplanes, which could contaminate the areas where these activities take place.

The means chosen by the farmer to "avoid contamination" was the creation of a buffer zone in the form of a contour pond about 4 metres wide and 2.5 metres deep, which surrounds the entire larger area of the dams; in the case of the smaller ponds, it was only created on the cultivation side, on the forest side it remained unobstructed.

However, the question arises as to whether the measures taken to prevent water and thus fish contamination are really effective, as most activities related to the application of products to plants are carried out by air.

On the other hand, it should be noted that the fish farmed in the company are destined for local and regional trade and even for export, so their water should be analysed by the bodies responsible for the quality of the products that reach the consumer. No information was provided on this.

The company "Delicious Fish", with an administrative centre in the district of Primavera in the municipality of Sorriso-MT, provides around 150 jobs directly linked to fish farming. It appears to be a successful company in the region, which has been featured in several reports in the national media as one of the national reference companies for freshwater fish farming with fish species from the Amazon region.

At the site where the fish breeding ponds are located, it was noted that these ponds do not impound the watercourses despite the large extent of the dams, which currently cover more than 170 ha of water, but it was noted that there was an elevation that may utilise water from the watercourse to supply the ponds during the dry season. Location with coordinates 12° 51'57.5" south latitude and 55° 51'57.5" west longitude, Figure 27. This practice is somewhat inconspicuous in the middle of the overgrown edge of the watercourse.

Figura 27 - **Pirarucu fish breeding tank**

Source: PERETTO, A. (25/11/2011).

THE PRODUCTION OF GEOGRAPHIC SPACE AND ITS EFFECTS

All human use of the geographical area leads to environmental changes, but some of them cause major environmental damage, resulting in enormous natural losses. The main environmental problems caused by the different uses in the Ribeirao Grande catchment area are: the construction of roads in headwater areas, leading to the death of vegetation; the construction of artificial drainage networks, which were discussed and analysed in the section on the characterisation of the channel valleys; and the problem of possible water pollution from the Delicious Fish farm, which was discussed in the section on land use.

However, a number of other problems have also been identified that have not yet been discussed and analysed. One of these is related to the deterioration of soil quality in the lower reaches of the catchment area.

A spring area was found and work is currently underway to restore it. This spring area is located in the lower reaches, where the spring valleys have steeper slopes and red latosols are present. [a]This valley is around 840 metres long and has a gradient of 40 metres, which is very steep for the region.

Based on field observations, this area is being restored because it is located on a steep slope, as shown in Figure 28, 12° 32' 06.5" south latitude and 55° 49' 33.1" west longitude, where the onset of mechanical erosion processes can be seen. The owner's interest in restoring the spring area is obvious, as mechanical erosion has begun in conjunction with stormwater runoff. This is one of the problems arising from the use of land with a slightly steeper slope, as was found at this site. It appears that the best use for this area is conservation and this was identified during the fieldwork.

Figura 28 - **A view of the valley where the land was prepared for reclamation, in a headwater area in the catchment area of the Ribeirao Grande.**

Source: PERETTO, A. (26/11/2011).

In the lower reaches of the Ribeirao Grande catchment area, visual and tactile analyses have shown that the soils have a sandier texture, which makes the erosive process, i.e. the formation of erosion in the soils of the crops, possible in the first place.

In Figure 29, at 12° 35' 23.8" south latitude and 55° 49' 05.5" west longitude, in the lower reaches of the basin, a ravine was formed. It is located on a red-yellow latosol soil with a sandy texture and corresponds to the headwaters of the Ribeirao Grande springs, a place without contour lines to control rainwater runoff. The ravine is located in a sloping area that drains the water to the spring area.

Figura 29 - **Erosion in a red/yellow latosol area**

Source: PERETTO, A. (2011).

Also of note in the lower reaches are isolated patches of vegetation that have not been cleared for agricultural planting, as there are outcrops of ferrous concretions, popularly known as gravel, which are currently used as road gravel and are being removed by landowners and authorities (Figure 30).

Figura 30 - **Machines removing gravel (location 12° 37' 20.2" S lat. and 55° 49' 37.4" W long., alt. 370 m).**
Figure 30 Erosion in a latosol area

Source: PERETTO, A. (27/11/2011).

In the Ribeirao Grande catchment area, an area used for highly mechanised agricultural production, producers are increasingly looking for techniques to increase their production.

Currently, 480 ha of maize are irrigated by 5 pivot systems, which is estimated to be 1.5 % of the utilised agricultural area. [33]To irrigate 480 ha with a 1 mm blade (set on the control panel of each pivot) would require 4,500 m of water or 4,500,000 litres (1 m = 1,000 litres). The 5 pivots are located in the middle and lower reaches sectors, the water for 4 pivots is taken from the Ribeirao Grande, which irrigates a total of 380 ha. Figure 31 shows the location of these pivots in the area in question. [a]A further area of 100 ha is irrigated with water from a 2nd order tributary of the Ribeirao Grande, which is located in the lower reaches.

Figura 31 - **Location of the irrigation points in the middle reaches of the Ribeirao Grande catchment area**

Pivos

dos pivos

Grande

4.7   9.3

3.3   6.6 km

W 55°56'   W 55°54'   W55°52'   W 55°50'   W 55°48'

S 12°34'   S 12°49'   S 12°43'   S 12°48'   S 12°53'

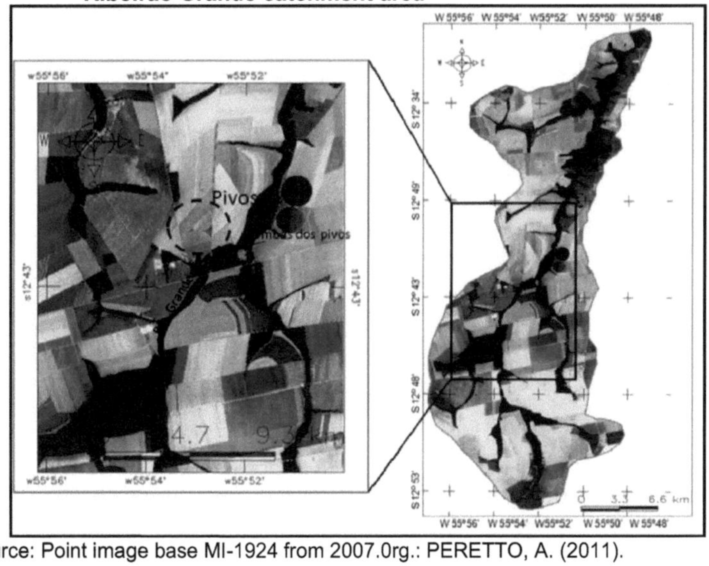

Source: Point image base MI-1924 from 2007.0rg.: PERETTO, A. (2011).

The amount of water that needs to be applied when irrigating crops depends on various factors, such as variety, cycle, planting time, planting density and climatic conditions at the planting site. Irrigation can be total (no rain) or supplemental (it rains on the plants but not enough to meet their needs).

The above data is an estimate to show how much water is taken daily from the rivers to irrigate these areas. If the irrigated areas increase, the water could no longer be available for this purpose and the Ribeirao Grande catchment area could face conflicts over the use of the water in its river system.

The pumps used to pump the water to irrigate the crops are installed on the banks of the Ribeiráo Grande and are powered by electricity. A suitable place for the pumps to collect the water was built in the river channel, as shown in Figure 32, which is located at the coordinates 12°42' 52.6" south latitude and 55° 53' 04.1" west longitude.

Figura 32 - **Where the water from the Ribeiráo Grande is collected for the irrigation systems**

Source: PERETTO, A. (25/11/2011).

The owners were asked for information to find out what requirements the environmental authorities have for the installation and operation of the standpipes, as this pump is used to irrigate 184 hectares of maize, which requires a lot of water. According to the owner of the standpipes, the SEMA (State Environmental Agency) required daily monitoring of the water flow until 2008, with two bathymetries being carried out to determine the potential flow of the rivers. [3]In 2008, in the month of August, the flow of the Ribeiráo Grande was 6 m /s. However, according to the landowner, the legislation changed and landowners were no longer obliged to carry out this type of control to determine the potential flow of the river, and SEMA was contracted to control the potential flow for the installation of taps for crop irrigation in the state of Mato Grosso.

This change in the law, which gives SEMA the authority to control the potential flow of rivers for the installation of pivot irrigation systems, could lead to problems in the future related to the application of this technique and cause an overload of water abstraction from the rivers in the catchment area. As public institutions are known to be bogged down in bureaucracy, more water could be abstracted than the natural flow potential of the rivers to be utilised for this purpose.

CHAPTER 6 CONCLUDING REMARKS

The Ribeirao Grande basin is an ideal reflection of the morphodynamic characteristics of the municipality of Sorriso. The investigation carried out as part of the study revealed that the basin has areas where the above-mentioned descriptions prevail, which, together with the climatic conditions, which have two clearly defined seasons: the dry season from May to September and the rainy season from October to April, become a set of factors that favour intensive agriculture.

The morphometric analysis of the Ribeirao Grande catchment showed that it has a low runoff density and flow density. However, it is a catchment with a high circularity index, a factor that shows that there is no probability of flooding associated with the main channel. However, there are areas of flooding at the tops of the confluences where the construction of artificial drains is concentrated to solve this type of problem, which hinders agricultural production in this basin and in other areas with these relief characteristics, with wide and very wide confluences.

aaAs far as the river hierarchy is concerned, it turned out to be a basin of order 4, with the rivers of order 1 being straight and the other channels of the other orders having alternating meandering and straight sections.

The geographical organisation of the Ribeirao Grande basin is geared towards agricultural activities, particularly the high level of technology used. The municipality of Sorriso currently has an excellent production per hectare for all crops grown.

The agricultural potential of the soil was of the utmost importance for the excellent level of production. These favourable conditions include the relief conditions provided by the Parecis plateau with its tabular and flat shapes and the wide and very wide incisions. These characteristics of the terrain have enabled mechanisation and made the use of machinery such as tractors and harvesters very efficient, as they do not have to manoeuvre over short distances, which reduces costs and speeds up production.

In relation to the relief, the importance of interfluves for the formation of latosols should also be emphasised, which, due to the easy correction of these soils, ensure the good productivity of crops in Sorriso and especially in the Ribeirao Grande basin. The predominant soil types in this basin are red, red-yellow and yellow latosols, together with hydromorphic soils.

In terms of land use in the Ribeirao Grande catchment area, 79% of the catchment area is used for agriculture. This category includes various traditional crops such as soya, maize and sorghum, and the activities that have recently emerged as alternatives in the region include fish farming and chicken farming.

The land use, the use of preserved vegetation, which corresponds to the areas with native vegetation in the catchment area (approximately 19.5%), shows that the catchment area almost fulfils the requirements of the Forest Code. There are areas where 20% of the area is preserved, which corresponds to the requirements for the Cerrado biome, and those where this is not the case and which need to be adapted. This adaptation of the APP reserves is likely because some countries that buy agricultural products from this region impose environmental requirements for the purchase of their products.

The vegetation of the basin is characterised by two different situations: the vegetation inside the valleys and the vegetation outside the valleys. In the valleys there is the emerging submontane forest and outside the valleys the vegetation of the open tree savannah and the dense tree savannah.

The development of the geographical area causes a number of environmental problems in the Ribeirao Grande catchment area: one of these is soil erosion, which occurs in some places. [a]Another factor is the opening of roads in first-order springs, where the dams have led to the death of vegetation due to flooding.

The solution to the problem of road construction in spring areas where the channel is not precisely defined would be to build roads around the springs and not within the spring valley.

It is extremely important that SEMA (State Environmental Secretariat), which is responsible for authorising the installation of irrigation taps for crops, withdraws the method of authorising the installation of taps so that in the future there are no serious problems in the use of water in the watersheds of the state of Mato Grosso, as landowners are not obliged to control the flow of the river to use it for crop irrigation. This fact that the flow is not controlled could lead to water stress in the watersheds of our state, since all water bodies have a capacity that can be withdrawn, and this capacity must be determined by bathymetry.

REFERENCES

FIND EVERYTHING AND THE REGION. history of Sorriso - MT. Available at: <http://www. achetudoeregiao.com.br/MT/sorriso/historia.htm>. Retrieved on: 10/06/2011.

ALMEIDA, F. F. M. de. 1948.º Reconhecimento geomOrfico nos planaltos divisores das Bacias Amazónica e do Prata, entre os meridianos 51 e 56o W Gr. Revista Brasileira de Geografía, 10 (3): 397 - 440, Rio de Janeiro.

ALVES, J. M. de P.; CASTRO, P. de T. A.; LANA, C. E. Influência de feigóes geológicas na morfologia da bacia do rio do Tanque MG. Revista Brasileira de Geociéncias, 33(2): p. 117-124, Sao Paulo, June 2003.

AMARAL, D. L.; FONZAR, B. C.; OLIVEIRA FILHO, L. C. de. Vegetagao. Phytoecological regions, their nature and economic resources. In: BRAZIL, Ministry of Mines and Energy. General Secretariat. RADAMBRASIL Project, Sheet SD.21/Cuiabá. PIN (Levantamento dos Recursos Naturais, 26). Rio de Janeiro: MME, 1982. pp. 401-452,

ARAÚJO, S. A.; KATO, S. L. R.; ROSESTOLATTO FILHO, A. O Grupo Bauru na Regiao de Poxoréu, e as Mineralizagoes Diamantíferas e suas Áreas Fontes. Area III Geology Study Report. Department of Mineral Resources. Institute of Exact and Earth Sciences. Federal University of Mato Grosso. Cuiabá: ICET/UFMT, 1991.

BARROS, A. M. et al. Geology. In: BRAZIL, Ministry of Mines and Energy. General Secretariat. RADAMBRASIL project, sheet SD.21/Cuiabá; geology, geomorphology, pedology, vegetation and potential land use. Rio de Janeiro: MME, 1982.

BERTALANFFY, L. VON. General systems theory. Petrópolis: Ed. Vozes, 1975.

BERTRAND, C. ; BERTRAND, G. Le Géosysteme: Un Espace-Temps Anthropisé. An ecological temporality. GEODE-UMR 5602, CNRS and University of Toulouse-Le Mirail. Toulouse, France : UTM, 1991.6 pp.

BERTRAND, G. Landscape and global physical geography. Methodological outline. "Révue Géographique des Pyrenées et du Sud Ouest". Translated by Dr Olga Cruz. Dr Olga Cruz. Caderno de Ciéncias da Terra, USP, n. 13, Sao Paulo, SP, 1971.

. Landscape and global physical geography. A methodological study. Révue Géographique des Pyrenées et du Sud Ouest, 39 (3): 249 - 27, Toulouse, France, 1968.

BITTENCOURT ROSA, D. Study of rocks with potential for crustal development in relief formation in the hydrographic basins of the Alto Paraguai, Juruena and Teles Pires rivers in the state of Mato Grosso. Final report for a post-doctoral fellowship. National Council for Scientific and Technological Development. Process No. 200.181/2004-1. Brasilia: CNPq, 2005.

BITTENCOURT ROSA, D.; GELA, A.; ALVES, D. de. O.; MACEDO, M.; GARCIA NETTO, L. da. R.; NASCIMENTO, L. A.; PINTO, S. D. S.; BORGES, C. A.; ROSSETO, O. C. TOCANTINS, N.; SANTOS, P. L. dos; GERALDO, A. C. H. Um Estudo Geoambiental Comparativo das Características Morfoestruturais e Morfoesculturais nas Áreas das Bacias do Alto Rio Paraguai e do Rio Teles Pires no Estado de Mato Grosso. Final report of the research project. Foundation for Research Promotion of the State of Mato Grosso. National Council for Scientific and Technological Development. Cuiabá: FAPEMAT/CNPq, 2002.

BRAZIL: National Environmental Council. °CONAMA Resolution No. 004, of 18 September 1985. Brasilia: D.O.U., 20 January 1986.

. Ministry of Agriculture, Livestock and Supply. Brazilian Society for Agricultural Research. National Soil Research Centre. Brazilian Soil Classification System. Brasilia: Embrapa Information Production; Rio de Janeiro: Embrapa Soils, 1999. 412 p.

. Ministry of Agriculture, Livestock and Supply. Brazilian Society for Agricultural Research. National Soil Research Centre. Brazilian Soil Classification System. 2nd revised edition. Rio de Janeiro: MAPA/EMBRAPA, 2006.

. The Presidency of the Republic. Law No. 9.433, of 8 January 1997, which establishes the National Water Resources Policy and creates the National Water Resources Management System. Brasilia: D.O.U. of 09/01/1997, section 1, p. 470, v. 135, n. 6.

BRAUN, E. H. G. Os Solos de Brasilia e suas Possibilidades de Aproveitamento Agricola. Revista de Geografia, (1), Rio de Janeiro, 1962.

CABRAL, I. de L. L.; JESUZ, C. R. de. Analysing relief-soil interactions as a tool to study the geographical distribution of latosols in the municipality of Campo Verde - MT. XVI National Conference of Geographers, Porto Alegre, 2011.

CABRAL, I. de L. L.; SILVA, G. F da. Effects of morphology on land cover in the municipality of Vera - MT. Geousp Magazine, v. 1, p. 36 - 50, Sao Paulo, 2011. '

CABRAL, T. L.; CABRAL, I. de L. L. Morphopedological approach as a tool for the study of the geographical distribution of latosols in the municipality of Sorriso - MT. Proceedings of the XVI National Meeting of Geographers - ENG, Porto Alegre, 2010.

CABRAL, T. L. Artificial drainage in wide catchments in the southern part of the municipality of Sorriso - MT. Thesis (Bachelor's degree in Geography). Department of Geography. Institute of Human and Social Sciences. Federal University of Mato Grosso. Cuiabá: UFMT, 2011.

CAMARA, G.; CASANOVA, M. A.; MAGALHAES, G. C.; MEDEIROS, C. M. B. Anatomy of Geographic Information Systems. Campinas-SP: UNICAMP, 1996.

CARDOSO, C. A.; DIAS, H. C. H. Morphometric characterisation of the Debossan catchment, Nova Friburgo - RJ. Forest Research Society. UFV. R. Árvore, v. 30, n. 2, p. 241-248, Vigosa-MG, 2006.

CHRISTOFOLETTI, A. Geomorfologia. 2nd edition. Sao Paulo: Edgard Blucher, 1980.

. Management proposals for sustainable development in micro-watersheds. Workshop of the Piracema Project, 2nd Nazaré Paulista-SP, Proceedings. Piracicaba-SP, 1969.

COLLARES, E. G. Evaluation of changes in the drainage networks of micro-catchments as a contribution to the geo-ecological zonation of river catchments: Application to the Capivari - SP catchment. Doctoral thesis. Sao Carlos School of Engineering. Sao Carlos-SP: USP, 2000.

CUNHA, S. B. da.; GUERRA, A. J. T. Environmental degradation. In: GUERRA, A. J. T.; CUNHA, S. B. da (Orgs.). Geomorphology and the environment. Rio de Janeiro: Ed. Bertrand Brasil, 1996. p. 337-379.

CUNHA, S. B. da. Fluvial geomorphology. In: GUERRA, A. J. T.; CUNHA, S. B. da (Orgs.). Geomorphology, an update of fundamentals and concepts. 3rd edition. Rio de Janeiro: Ed. Bertrand Brasil, 2000.

DERBY, O. A. Nota sobre a Geologia e a Paleontologia de Mato Grosso. Archives of the National Museum, (9): 59-88, Rio de Janeiro, 1895.

DURAND-DASTÉS, F. Climatology. Encyclopaedia Universalis, 4, pp. 618-624, Paris, France, 1968.

ESTIENNE P. and GODARD. A. Climatology. Armand Colin. Paris, 1970 .

FERREIRA, J. C. V.; SILVA, Pe. J. de M. Cidades de Mato Grosso. Cuiabá: J. C. V. Ferreira, 2008.

FONSECA, G. P. S. Analysis of diffuse pollution in the Teles Pires catchment using geoprocessing techniques. Master dissertation. Federal University of the State of Mato Grosso. Cuiabá - Mato Grosso. 2006.

FREITAS, R. O. Drainage texture and its geomorphological application. Boletim Paulistade Geografia, v.11, pp. 53-57, Sao Paulo, 1952.

GUERRA, A. J. T.; CUNHA, S. B. da. (Orgs.). Geomorphology and the environment, Rio de Janeiro: Ed. Bertrand Brasil, 1996. 394 p.

HORTON R. E. Erosive evolution of streams and their drainage base: hydrophysical approach to quantitative morphology. Geol. Soc. America Bulletin, 56 (3), pp. 275-370, New York, USA, 1945.

IBGE. Brazilian Institute of Geography and Statistics. IBGE Cidades@. Census 2010. available at: <http://www.ibge.gov.br/cidadesat/topwindow.htm21>. Retrieved on: 15/06/2011.

. Brazilian Institute of Geography and Statistics. Technical Manuals of Geosciences. Rio de Janeiro: IBGE, 2007.

. Brazilian Institute of Geography and Statistics. Technical Manual on Land Use. Rio de Janeiro: IBGE, 1999. 58 pp. (Technical manuals in geosciences, no. 9).

JUSTINIANO, L. A. A. Fluvial dynamics of the Paraguay River between the mouth of the Sepotuba and the mouth of the Cabagal. Master dissertation. Postgraduate Programme in Environmental Sciences. State University of Mato Grosso. Cáceres-MT: UNEMAT. 2010.

KER, J. C.; PEREIRA, N. R.; CARVALHO JUNIOR, W. de; CARVALHO FILHO, A. de. Cerrados: Soils, suitability and agricultural potential. Simposio sobre Manejo e Conservagáo do Solo no Cerrado, Ed. Fundagao Cargill, pp. 1-19, Goiania, GO, 1990.

MACIEL, M. A. C. & RIBEIRO, J. M. C. Contribuigao ao Estudo do Grupo Bauru, Unidades Terciarias e Quaternárias e o Diamante na Regiáo de Poxoréu, MT. Area II Geology Final Report, Department of Mineral Resources, Institute of Exact and Geosciences, Federal University of Mato Grosso, 65 p., Cuiabá, MT. 1991.

MELO, D. P. de; FRANCO, M. do S.M. Geomorphology. In: Brazil. Ministry of Mines and Energy. General Secretariat. Project RADAMBRASIL.

sheet
SC.21 Juruena. Rio de Janeiro, Radambrasil. 1980 (Levantamento de Recursos Naturais, 20).

MENDES, J. C. Elementos de Estratigrafía. Sao Paulo: ed. T. A. Queiroz, 1996 (Biblioteca de Ciéncias Naturais).

MONTEIRO, C. A. F. Geosistemas: a história de uma procura. Sao Paulo: Contexto Publishing House, 2000.

MOREIRA, M. L. C.; VASCONCELOS, T. N. N. Mato Grosso SOLOS e

PAISAGENS. Cuiabá: Ed. Entrelinhas; SEPLAN-MT, 2007. .

______________ .... _ ...
______________ MOREIRA, R. O que é geografia. Sao Paulo: Editora Brasiliense, 2005. 113 p.

MORENO, G. Agriculture: Transformations and trends. In: MORENO, G.; HIGA, T. C. C. de S. (Orgs.); MAITELLI, G. T. (Collab.). Geography of Mato Grosso: territory, society, environment. Cuiabá: Entrelinhas, 2005. p. 140-171.

MÜLLER, V. C. A quantitative geomorphology study of drainage basin characteristic in the Clinch Mountain Area. New York: Virginia and Tennesse. Dept. of Geology. n. 3, p. 30, 1953.

NETO, A. L. C. Slope hydrology at the interface with geomorphology. In: CUNHA, S. B; GUERRA, A. J. T. (Orgs.) Geomorfologia: uma actualizagao de bases e conceitos. 4th ed. Rio de Janeiro: Editora Bertrand Brasil, 2001. p. 93-148.

OLIVEIRA, A. L. A. M. de. The Bauru Group and the Tertiary and Quaternary units from Poxoréu to General Carneiro as carriers of diamond mineralisation and their source areas. Final research report. Institutional programme for scientific initiative grants. Federal University of Mato Grosso. National Council for Scientific and Technological Development. Cuiabá: PIBIC/UFMT/CNPq, 1992.

OLIVEIRA, A. L. A. M. de.; COSTA, T. A. and BARBOSA, C. R. da. R. The clastic-volcanic-chemical sequence of the Bauru Group and the Tertiary and Quaternary units of the Dom Aquino region - MT. Geology final report. Department of Mineral Resources. Institute of Exact and Earth Sciences. Federal University of Mato Grosso. Cuiabá: ICET/UFMT, 1992.

OLIVEIRA, V. A. de; AMARAL FILHO, Z. P.; VIEIRA, P. C.. Soil science. In: BRAZIL, Ministry of Mines and Energy. General Secretariat. RADAMBRASIL Project; Sheet SD.21/Cuiabá; Geology, geomorphology, pedology, vegetation and potential land use. Rio de Janeiro: MME, 1982.

OLIVEIRA, V. A. de; AMARAL FILHO, Z. P.; VIEIRA, P. C. 1982. soil science. In: BRAZIL, Ministry of Mines and Energy. General Secretariat. RADAMBRASIL project, sheet SD.21/Cuiabá. PIN (Levantamento dos Recursos Naturais, 27). Rio de Janeiro: MME, 1982. pp. 257-400.

PERETTO, A. ; SOUZA, C. A. . Córrego Santissimo Hydrographic Basin: Geoenvironmental Aspects and Land Use. Journal Geopantanal, v. 5, p. 141-162,

PIAIA, I. I. Geography of Mato Grosso. 3rd ed. rev. ampl. Cuiabá: EUNIC, 2003.

PISANI, J. R. T.; ARRAIS, J. C. de P. Contribuigao ao Estudo do Grupo Bauru, Unidades Terciarias e Quaternárias e o Diamante na Regiao de Poxoréu, MT. Area I. Geology final report. Department of Mineral Resources. Cuiabá: ICET/UFMT, 1991.

ROCHA, C. H. B. Geoprocessing: transdisciplinary technology. Juiz de Fora: Ed. do Autor, 2000.

ROSS, J. L. S.; SANTOS, L. M. dos. Geomorphology. In: BRAZIL, Ministry of Mines and Energy. General Secretariat. RADAMBRASIL Project, Sheet SD.21/Cuiabá. PIN (Levantamento dos Recursos Naturais, 26). Rio de Janeiro: MME, 1982a. S. 193-256.

. Geomorphology. In: BRAZIL, Ministry of Mines and Energy. General Secretariat. RADAMBRASIL project, sheet SD.21/Cuiabá; geology, geomorphology, soil science, vegetation and potential land use. Rio de Janeiro: MME, 1982b.

ROSS, J. L. S. Ecogeography of Brazil: subsidies for environmental planning. Sao Paulo: Oficina de Texto Publishing House, 2009.

. Geomorphology: Environment and planning. 6th edition. Sao Paulo: Editora Contexto, 2001. 85 p.

...

SANTOS, Eduardo Vieira dos et al. THE OCCUPATION OF THE CERRADO BIOME: from the expansion of the agricultural frontier to the present day. Vil SIMPOSIO DE HISTORIA - CLIO E SEUS ARTÍFICES, Rethinking the Origin of History. Federal University of Goiás. Campus de Catalao-GO, 2004.

SOTCHAVA, V. B. The Study of Geosystems, 16 Methods in Question. Sao Paulo: USP/Department of Geography, 1977.

SORRISO. The town hall. Features. Available at: <http://www.sorriso.mt.gov.br/>. Accessed on: 08/06/2011.

SPIRONELLO, R. L.; DE BIASI, M. Avaliação de conflitos ambientais na microbacia do Arroio Taquarussu - municipio de Ipora do Oeste/SC. GEOUSP - Espago e Tempo, n. 17, pp. 61-79, Sao Paulo, 2005.

TARDY, Y. Le Cycle de L'Eau - Climats, Paléoclimats et Géochimie Globale. Paris, France: Masson Editeurs, 1986.

TRICART, J. Ecodynamic classification of the environment. In: Ecodynamics. Rio de Janeiro: Fibge, 1977.

TUCCI, C. E. M. Flooding in cities. In: TUCCI, C. E. M.; PORTO. R. L. L.; BARROS, T. de B. (Orgs). Urban drainage. Chapter 1. Porto Alegre: ABRH/Editora da UFRGS, 1995. p. 15-31.

VILLELA, S. M.; MATTOS, A. Hidrologia aplicada. Sao Paulo: McGraw-Hill do Brasil, 1980.

WESKA, R. K. Diamond Placers from the Água Fria Region, Chapada dos Guimaraes - MT. Master dissertation. Institute of Geosciences. University of Brasilia. Brasilia: UnB, 1987. 170 p.

. A synthesis of the Upper Cretaceous of Mato Grosso. Rev. Geociencias, v. 5, n. 1, p. 71-81, UNESP, Sao Paulo, SP, 2006.

WESKA, R. K.; SVISERO, D. P.; LEONARDOS, O. H. Contribuigao ao Conhecimento do Grupo Bauru no Estado de Mato Grosso, Brasil. In: Boletim do Simpósio sobre o Cretáceo do Brasil, 4, pp. 289-295, UNESP, Rio Claro, SP, 1996.

Table of contents

I want morebooks!

Buy your books fast and straightforward online - at one of world's fastest growing online book stores! Environmentally sound due to Print-on-Demand technologies.

Buy your books online at
www.morebooks.shop

Kaufen Sie Ihre Bücher schnell und unkompliziert online – auf einer der am schnellsten wachsenden Buchhandelsplattformen weltweit! Dank Print-On-Demand umwelt- und ressourcenschonend produziert.

Bücher schneller online kaufen
www.morebooks.shop

Printed by Books on Demand GmbH, Norderstedt / Germany